PROBLEM OF LAND SNAIL PESTS IN AGRICULTURE

ABOUT THE AUTHOR

Born at Azamgarh (Uttar Pradesh), the author had his schooling in his home town after which he moved to Allahabad for higher education. After doing his B.Sc., M.Sc., and D.Phil. from the University of Allahabad, he joined the Indian Agricultural Research Institute, New Delhi (IARI) where he did both teaching and research. At IARI, for a few years, he worked on Insect Physiology and Locust Toxicology. However, he had to change his line of work in 1965 and started working on land snail pests especially the giant African snail (the biggest-sized land snail in the world) both at IARI, New Delhi and on the Andamans. He continued to work on this pest till his retirement in December 1980. This work has been widely appreciated and quoted by authorities like Prof. A.R. Mead of the University of Arizona, U.S.A.

The author was selected as a member of Post-Graduate Faculty, from the very inception of the Post-Graduate School at IARI in 1958. He took M.Sc. and Ph.D classes and guided a number of students leading to Master and Doctorate degrees.

In addition to having 50 research papers to his credit, published in Indian and Foreign journals, he has also authored a book *Economic Zoology* which is now being revised. He has also authored another book *Faslon Ke Shatru—Bharat Ke Thal Ghonghe* (in Hindi). It is being published by ICAR, New Delhi.

He is one of the editors of *Agricultural Entomology* (Vols. I and II). He has also edited another book *Crops and Weather* as an External Editor and it is being published by ICAR, New Delhi.

The author is a Fellow of the Entomological Society of India, New Delhi and Life Member of (i) National Academy of Sciences, Allahabad and (ii) All India Scientific Writers' Society, New Delhi.

He is also recipient of one of the 1984 *Statesman Awards for Rural Reporting.*

PROBLEM OF LAND SNAIL PESTS IN AGRICULTURE

(A Study of the Giant African Snail)

P.D. SRIVASTAVA
Former Senior Scientist-Entomologist
Indian Agricultural Research Institute, New Delhi

Foreword by
M.S. Swaminathan

CONCEPT PUBLISHING COMPANY, NEW DELHI-110059

ISBN 81-7022-435-7

First Published 1992

Published and Printed by
Ashok Kumar Mittal
Concept Publishing Company
A/15-16, Commercial Block, Mohan Garden
NEW DELHI-110059 (India)

Lasertypeset by
Microtech Advance Printing Systems (Pvt.) Ltd.
H-13, Bali Nagar,
NEW DELHI-110015

FOREWORD

Snail pests cause considerable damage to food crops and particularly to high value horticulture crops. Sometimes these are accidentally introduced as happened some years ago in the case of the Giant African Snail in Andaman Islands where they soon became serious pests.

Dr. P.D. Srivastava and Scientists of the Entomology Division of IARI, New Delhi have carried out a comprehensive study of the biological and management aspects of snail pests with particular reference to the Giant African Snail. This book is hence an authoritative treatise on this subject. I am confident, it will be found valuable by scientists, research scholars, extension workers and farmers. We owe a deep debt of gratitude to Dr. Srivastava for this labour of love.

M.S. SWAMINATHAN

PREFACE

Achatina fulica Bowdich is rightly known as the giant African snail because of its large size and African nativity. It is the biggest-sized and economically the most important land snail pest hitherto known in the world. Although it is not a native of India but Gude (1914) considers it "advisable to include it (in the Fauna of British India) since it appears to have established itself thoroughly (in India)". It is a voracious feeder and causes huge economic losses to a large number of crops in the "tropical and subtropical regions" of the world. In India, this snail causes severe damage to a number of agricultural especially the horticultural and plantation crops mainly in Bihar, Assam, West Bengal, Orissa, Kerala and Tamil Nadu, in addition to a number of islands of the Andaman and Nicobar group, so much so that certain cultivators have stopped growing vegetables. So far, there does not appear to be any consolidated published account dealing with different aspects of *A. fulica* from the Indian subcontinent in any language; the only available published account has been authored jointly by Prof. K.C. Ghose and Dr. S.K. Raut of the University of Calcutta, dealing with certain aspects of seven land snails including *A. fulica.* Prof. Ghose has made valuable contribution to the knowledge especially of the anatomy and morphology of this snail. Amongst the foreign publications, the noteworthy publication is the book on giant African snail authored by the well known Malacologist, Prof. A.R. Mead of the University of Arizona, U.S.A. However, even he has not dealt with most of the fundamental aspects. In the present publication, attempt has been made to present a critical review of the available information on practically all aspects of this pest.

I have worked on the giant African snail for over sixteen years at the Indian Agricultural Research Institute, New Delhi and its substation at Port Blair. Laboratory tests were performed mainly at Delhi and field trials were held on some islands of the Andaman and Nicobar group especially on the main island of South Andaman.

An attempt has been made to present a consolidated account of up to date information available on this important snail pest which continues to remain a problem for the cultivators who are unable to grow crops of their choice and have to change the cropping patterns to save their crops from the ravages of this pest. Special effort has been made to present a detailed and practically feasible account of the management of this pest. In the end, lacunae have been identified and future lines of research suggested.

I hope, students, teachers, researchers and cultivators, all will find it useful. If this hope is fulfilled I shall consider myself amply rewarded.

New Delhi

P.D. SRIVASTAVA

ACKNOWLEDGEMENTS

I wish to express my deep sense of gratitude to the Department of Science and Technology (DST), Government of India, New Delhi for giving full financial support for twenty months to the project "*Achatina fulica* Bowdich, the giant African snail — A monograph" and to the National Academy of Sciences, Allahabad for forwarding the project to the DST and for looking after the financial aspects including maintenance of accounts. The General Secretaries and the Executive Secretary of the Academy need special mention for their helpful and cooperative attitude throughout.

I am grateful to Drs. G. Nath, K.R. Gupta, Anuj Sinha, Ishraq Ahmad and Shri B.P. Singh of DST for their prompt help and advice whenever needed. I am equally thankful to my former colleagues Drs. G.P. Gupta, K.L. Doharey and Y.N.S. Srivastava of the Indian Agricultural Research Institute (IARI), New Delhi for giving occasional help and advice as willingly as I used to get when they worked with me at the Institute. My special thanks are also due to Prof. U.S. Srivastava, formerly Head of the Department of Zoology, University of Allahabad. He helped me at all the stages of the preparation of the monograph especially in the finalisation of Chapter 5 dealing with morphological aspects.

I am also thankful to :

Dr. Jagdeep Saxena of the Editorial Board of the Indian Council of Agricultural Research (ICAR) and Dr. Saroj Tuli of Maitriya College of the University of Delhi, for giving valuable suggestions for the improvement of the manuscript.

Dr. A.K. Bandopadhyaya, Director, Central Agricultural Research Institute, Port Blair for giving facilities and to Sarvashri Prashanth and H.R. Ranganath of the same Institute for cooperating during the course of my visit to Port Blair for assessing the latest position of *A. fulica*.

Prof. S.F. Hameed, Head, Department of Entomology, Rajendra

Agricultural University, Pusa (Bihar) and Dr. P.N. Narula, Head, IARI-Regional Station, Pusa for giving all facilities during my visit to Pusa (Bihar) to assess the position of *A. fulica.*

Head, Library Services, IARI Library and Incharge Library of the Division of Entomology, IARI, for according permission to consult literature in the Library.

Prof. K.C. Ghose, formerly Professor of Zoology, University of Calcutta and Dr. S.K. Raut of the same University for sending to me reprints and photocopies of their research papers on *A. fulica* on account of which the task of writing the monograph became easier.

Dr. N.V. Subba Rao and Dr. V.D. Srivastava of the Zoological Survey of India, Calcutta for sending to me their reprints related to the topic.

Dr. F.J. Olson of the Plant Protection and Quarantine Programmes, U.S. Department of Agriculture, Honolulu, Hawaii for sending to me even his unpublished list of host plants of *A. fulica.*

Prof. R.B. Dumbre, Head of the Department of Entomology of the Konkan Krishi Vidyapeeth, Dapoli (Maharashtra) for giving me the information regarding the presence of *Achatina* in the Konkan region of Maharashtra by sending to me live specimens of the snail.

Late Prof. T.A. Davis, Director, J.B.S. Haldane Research Centre, Nagercoil (Tamil Nadu) for sending to me photographs of the coconut crab — a predator of *Achatina.*

P.D. SRIVASTAVA

CONTENTS

LIST OF FIGURES

LIST OF TABLES

1

INTRODUCTION

Achatina fulica Bowdich (Fig. 1.1), the giant African snail, *is economically the most important land snail pest* in the world. Although originally of Madagascar/East Africa, it has reached many countries of the world, including India. It was introduced by W.H. Benson in Calcutta in 1847. In spite of the existence of quarantine regulations in different countries, the "snails are continuing to spread faster than more efficacious control measures are being discovered" (Mead, 1979). In spite of the "excellent quarantine regulations of long standing in Australia", *A. fulica* has managed to reach the North-East coast of Queensland (Australia) and was located there in April 1977 (Mead, 1979).

According to Mead (1979) "the story of Economic Malacology is essentially the story of the giant African snail" which has been spreading from its original homeland to different countries of the world since nineteenth century. This snail is a serious pest of fruit, vegetable and plantation crops as well as of a number of ornamentals. Mead (1961), Olson (unpublished) and Raut and Ghose (1984) have given long lists of its host plants. Another indication of its importance can be judged by the fact that in "successful campaign in Florida" (USA) against *A. fulica,* more than $ 300,000 were spent in collection and destruction of "over 18,000 snail specimens and thousands of eggs" (Mead, 1979).

Taxonomically *A. fulica* belongs to class Gastropoda of Phylum Mollusca. Molluscs are soft-bodied unsegmented animals, typically having an anterior head, a ventral foot and a dorsal visceral mass. The body is covered with a thin fleshy mantle which secretes a protective calcareous shell. The shell may be internal, reduced or even absent. Molluscs have been highly successful animals both in terms of the number of individuals and the number of species (Russel-Hunter, 1979).

Molluscs are mainly marine. Certain snails and some pelecypods occur in brakish and fresh water also; other snails and slugs are terrestrial (Storer and Usinger, 1965).

FIG. 1.1: *A. fulica* — parts of body protruding out of the shell

Different estimates regarding their numbers have been given by different authors but all agree that Phylum Mollusca stands second only to Phylum Arthropoda with regard to the number of living species.

According to Storer and Usinger (1965) the number of living species of molluscs is 45,000 whereas Wright (1971) estimates the number to be 110,000 mostly belonging to the classes Gastropoda, Pelecypoda and Cephalopoda. Russel-Hunter (1979) states that "more than 99 per cent of living molluscan species are snails and bivalves." Wright and Russel-Hunter both state that the estimated number of living species of molluscs is more than twice the total number of all known vertebrate species even if the lower figure (i.e., 80,000) is considered as the actual number of living molluscs.

Classification

Phylum Mollusca is divided into six classes : (1) Monoplacophora, (2) Amphineura, (3) Scaphopoda, (4) Cephalopoda, (5) Pelecypoda (Bivalvia) and (6) Gastropoda.

1. *Monoplacophora*

The entire class was considered to have fossilised till, amongst the molluscs collected by the Zoologists of the Danish Galathea Expedition in 1952, specimens of *Neopilina galatheae* were found (Johnson *et al.*, 1969) and were collected from the dark muddy clay at a depth of 3350 m from the Pacific Ocean, off the coast of Costa Rica (Storer and Usinger, 1965). *N. galatheae* measured 37 × 33 mm (Kaestner, 1967). In 1958, Zoologists of the American Research Vessel Vema discovered another species *Neopilina* (Vema) *ewingi.* This also was dredged out from the Pacific at a depth of about 5850 m near Lima (Kaestner, 1967; Storer and Usinger, 1965). *Neopilina* is a deep water marine form. Living members of this class are bilaterally symmetrical and single shelled. This genus is unique in having segmented body and many of its organs are paired (Wright, 1971).

Neopilina has five to six pairs of gills, six pairs of nephridia, and a flat, almost circular foot. This class as a whole is unisexual and has no economic importance.

2. *Amphineura*

All members of this class also are marine. This class includes Chitons. Body is oval or dorsoventrally compressed and long. These are generally 2-3 cm long but some may be even longer. These have many external gills and a single pair of nephridia. Shell in a Chiton is made up of 8 broad plates. In the event of danger, they can roll upon themselves and assume the shape of a ball. According to Storer and Usinger (1965) the number of living species of this class is 700, whereas Wright (1971) puts it only as 500. These also are unisexual and have hardly any economic importance.

3. *Scaphopoda*

These are commonly known as tooth shells or tusk shells. All are marine comparatively smaller, bilaterally symmetrical animals where mantle and shell are tubular and slightly curved; shell opens at both the ends. These are bottom-living molluscs and are found from shallow waters up to a depth of 4600 metres (15,000 ft). Head and gills are absent. Tentacles are present around the mouth. Respiration takes place through mantle. Like Monoplacophora and Amphineura these are also unisexual. The number of living species is 200 (Storer and Usinger, 1965), whereas Wright (1971) puts this number as 300. Their shells have some economic importance as articles of show or beauty.

4. *Cephalopoda*

This class also is exclusively marine and includes *Nautilus, Loligo, Sepia,* squids and octopuses. These are most highly specialised molluscs. This class includes the largest living invertebrates having well developed eyes on the head. These are highly mobile and are carnivorous. Shell may be external, internal or absent. Some forms have chambered-external shell. Head is surrounded by 8-10 arms generally having suckers at their free ends. The arms are actually modified form of the foot. These have one or two pairs of well developed gills. These are unisexual. The number of living species of the class has been reported as 400 by Storer and Usinger (1965) as well as by Wright (1971). This class is economically quite important.

5. *Pelecypoda (Bivalvia)*

This class has marine as well as fresh water forms and includes fresh-water mussels or clams and maine forms like clams, *Teredo* (Shipworm), oysters etc. Because of their filter-feeding habit, they are confined, exclusively to aquatic habitat (Wright, 1971). Body is enclosed in a bivalved shell, the two valves being joined together by a dorsal hinge. They are bilaterally symmetrical, the mantle has right and left lobes. There are no jaws and head. Foot is hatchet-shaped and extends between the valves during movement. One or two pairs of gills are present. Like the four classes of Mollusca, described earlier, these also are unisexual. Storer and Usinger (1965), put the number of living species of Pelecypoda as 11,000 whereas Wright (1971) puts it as

20,000. Numerically, amongst the molluscan classes it stands second only to Gastropoda (Wright, 1971). Pelecypods are of immense economic importance because of being edible and 'manufacturers' of genuine pearls.

6. *Gastropoda*

This is numerically, the largest class of Phylum Mollusca, the living species numbering 33,000 (Storer and Usinger, 1965). Its members have successfully adapted to different habitats, that is why they are found in sea, fresh water and on land. Their feeding habits range from browsing to sedentary, filter feeding to carnivorous and predatory. The class includes *Cypraea,* conch shell, snails and slugs. According to Russel-Hunter (1979), more than 99 per cent of living species of molluscs are snails and bivalves.

These are asymmetrical and the body lies within a one-piece spirally coiled shell. In slugs and some others external shell may be absent or when present, may be as a rudimentary internal thin plate. The body has a distinct head, tentacles and eyes. Ventral part of their foot is flat with which they crawl. This class has commercial importance, in addition to being important in health and agriculture.

This class is divided into three sub-classes : (i) subclass *Prosobranchia* : These are mostly marine and have a single pair of tentacles. These are unisexual with a single sex opening. This subclass is the largest of the three gastropod subclasses, having 25,000 living species including *Cypraea moneta* (Cowries) and the conch shell, (ii) subclass *Opisthobranchia:* These are exclusively marine and bisexual shell is absent or reduced. These have two pairs of tentacles. The number of living species of this subclass is 3000 and (iii) subclass *Pulmonata* includes aquatic and terrestrial snails and slugs. Shell may be simple or spirally coiled into which the snail can withdraw or it may even be absent. In the case where there is no external shell, there may be an internal plate which is actually a rudiment of shell. Tentacles may be one or two pairs; ctenidia are absent. Mantle cavity is anterior in position and is transformed into a pulmonary sac with narrow pore the pneumostome on the right side anteriorly. There are no gills and respiration takes place through the pulmonary sac. It is filled with air by the snail's rising to the surface and opening of the pneumostome through the surface film. Head is usually not well-demarcated from the body, but in some, there is groove below the mouth and which marks anterior border of the foot (Hyman, 1967). These are all bisexual. The number of living species is more than 5000.

Subclass pulmonata is divided into two orders :

(i) Order *Basommatophora* members of this order have a single pair of invaginable tentacles, near the base of which are situated the eyes. Male and female gonophores are generally separate. These are all aquatic, generally fresh-water forms e.g., *Lymnaea* and *Planorbis,* but some forms are marine also (Parker *et al.,* 1956). According to Wright (1971) members of this order are "mostly fresh-water snails with a few terrestrial and some marine forms".

(ii) Order *Stylommatophora* is exclusively terrestrial, includes land snails and slugs. Members of this order have two pairs of invaginable tentacles; at the tip of the posterior pair are situated the eyes. Both the pairs of tentacles are sensitive to touch. This order has great agricultural importance as this includes land snail and slug pests of agriculture and also predatory snails which feed upon land snail pests. *A. fulica* also belongs to this order. All land pulmonates share adaptations for maintaining internal water in aerial environment. The pulmonate slugs which have secondarily lost their shells differ only in coping with extreme water loss and in behavioural mechanisms which prevent their desiccation and death (Russel-Hunter, 1979). Skin of land pulmonates is readily permeable to water. Snails which are shelled, however, can greatly cut down their water losses by withdrawing in the shell and closing the opening of the shell by their secretion. *Most of the species are nocturnal and take advantage of cooler night temperatures and the presence of free water on plant surfaces in the form of dew.* Fertilization is reciprocal (Russel-Hunter, 1979). Thus each mated land pulmonate is capable of laying eggs; male and female gonophores are usually united. This provides greater surity for the survival of the species of this order.

Taxonomic position of *Achatina fulica* Bowdich is given below :

Phylum	—	Mollusca
Class	—	Gastropoda
Subclass	—	Pulmonata

Order	— Stylommatophora
Family	— Achatinidae
Genus	— *Achatina*
Species	— *fulica*

Some other Land Snail Pests of India

In addition to *Achatina fulica,* undoubtedly recognised as economically the most important land snail pest, all over the world, there are a number of other land snails occurring as pests in India as listed below along with their host plants.

As already stated earlier, all land snails belong to the pulmonate order Stylommatophora.

Two important land snail pests, *Opeas gracile* (Hutton) and *Zootecus insularis* (Ehrenberg) belong to the family Subulinidae.

(1) *Opeas gracile* is a tiny international land snail pest (shell-size generally, 13.5 × 3.5 mm) distributed all over the world except Australia in the tropical and sub-tropical regions (Gude, 1914). It is generally a pest of green house and other potted plants. Ray and Mukherjee (1963) report that in addition to green house and potted plants, it also causes damage to other garden plants.

(2) *Zootecus insularis* is slightly larger than *O. gracile* in size. Largest shell-size reported from India is 13.8 × 5 mm (Subba Rao and Mitra, 1979). Its distribution in India has been recorded by Gude (1914) from Delhi, Uttar Pradesh, Bihar, Gujarat, Maharashtra, Kashmir, Tiruchirapalli. Subba Rao and Mitra (1979) have recorded it from Andhra Pradesh, Maharashtra, Gujarat, Rajasthan, Punjab, Kashmir, Madhya Pradesh, Uttar Pradesh and Bihar. The author has observed this snail in Delhi as a pest of okra (*Abelmoschus esculentus*) and mint (*Mentha arvensis*).

Family Ariophantidae has four important genera namely *Ariophanta, Cryptozona, Macrochlamys* and *Bensonia.* Some species of these are pests. *Ariophanta* has three important species : (i) *A. bajadera* (Pfeiffer) (ii) *A. solata* (Benson) and (iii) *A. ligulata* Ferrusc.

(3) *A. bajadera* is distributed mostly in Gujarat and Maharashtra. Within the past five years, this snail has assumed the form of a serious pest and has been observed causing damage to okra, money plant, shoe flower (China rose), groundnut, seedlings of jack fruit and teak. The most seriously affected crop in Konkan region of Maharashtra is groundnut (Personal communication, R.B. Dumbre, 13.7.88).

(4) *A. solata* appears to be restricted only to the southern region, especially parts of Tamil Nadu, Andhra Pradesh, Kerala and Karnataka (Raut and Ghose, 1984).

This snail causes damage to coffee (*Coffea arabica*) and Coral tree (*Erythrina variegata*). The tender pods and shoots are edible and leaves are fed to cattles. The snails feed upon the green parts of the tender stems, leaves and also green berries. The tender stems fed upon by the snail break when exposed to strong winds.

(5) *A. ligulata* is distributed in the eastern parts of the peninsular region, parts of Karnataka, Tamil Nadu, Orissa and Bihar.

It is a pest of vegetables and ornamentals. Detailed information in this regard is not available.

Three species of genus *Cryptozona* i.e., (i) *C. (Xestina) bistrialis* Beck, (ii) *C. (Nilgiria) semirugata* (Beck) and (iii) *C. belangiri* Deshayes are fairly common pests of horticultural crops mainly in the peninsular region of the country.

(6) *C. (Xestina) bistrialis* is distributed in certain parts of Andhra Pradesh, Tamil Nadu, Karnataka, Kerala and Maharashtra. It feeds upon the tender parts of rubber plants mainly in the nursery (Subba Rao, 1975). This snail also drinks latex from the apical portions by lacerating tender buds and while climbing down drinks latex from the tapping spot (Pillai and Jayarathnam, 1983).

(7) *C. (Nilgiria) semirugata* is common in certain parts of Gujarat, Tamil Nadu and West Bengal. It has been observed feeding on the leaves of Oleander (*Thevetia* sp.) and on an evergreen shrub *Adhatoda vasica* of medicinal importance (Subba Rao, 1975). It also feeds upon the leaves of another medicinal plant, *Solanum torvum* (Laxminarayana and Subba Rao, 1971).

(8) *C. belangiri* is distributed in the southern region and Kolhapur (Maharashtra). It has been reported by Jagtap *et al.*, (1989) to be a pest of groundnut, beans, green gram (*Vigna radiata*), black gram (*V. mungo*), chillies (*Capsicum annuum*), sunflower (*Helianthus annus*), grapes (*Vitis vinifera*), betelpepper (*Piper betle*) and turmeric (*Curcuma domestica*) in the moist areas of submontane zone of Maharashtra during the rainy season.

As regards the genus *Macrochlamys*, about 84 species of this genus are distributed in the coastal, hilly regions and plains of the country. However, adequate information regarding food plants of most of these is not available except of *M. indica* (Blanford and Godwin-Austen 1908).

(9) *M. indica* is the common garden snail reported from certain parts of Assam, Bihar, Orissa, West Bengal, Kerala and Tamil Nadu (Raut and Ghose, 1984). The author has observed the presence of this snail in Delhi, Pusa (Bihar) and Port Blair. It is a pest of Cucurbitaceous plants e.g., *Luffa* spp. beans, lettuce, cole crops, drum stick (*Moringa oleifera*), amaranth (*Amaranthus blitum*), okra (*Abelmoschus esculentus*), marigold (*Tagetes* spp.) and *Chrysanthemum* sp. (Raut and Ghose, 1984). The author has observed this snail feeding upon money plant (*Scindapsus (Pothos) aureus*) also.

(10) There is only one species of genus *Bensonia* i.e., *B. monticola* Hutton — a land snail of mountain regions. It is distributed in Sikkim, Himachal Pradesh, Jammu and Kashmir, Kumaon, Garhwal and Mussoorie in Uttar Pradesh (Blanford and Godwin-Austin, 1908; Raut and Ghose, 1984). It has been reported to practically defoliate bean plants. It also feeds upon Ipomea and Okra (Bhardwaj and Thakur, 1973). Bhalla and Pawar (1977) have reported this snail to be a pest of a number of fruit crops like papaya (*Carica papaya*) and *Citrus* spp. etc.

2

DISPERSAL/DISTRIBUTION OF THE GIANT AFRICAN SNAIL IN THE WORLD[1]

Mead (1949) has referred to *A. fulica* as a native of East coast of Africa, from Natal to Mozambique in the south and to Kenya and Somaliland in the north. Gude (1914) in the "Fauna of British India" has referred to *A. fulica* as a native of Madagascar — an East African Island situated about 400 km across Mozambique Channel. Quite a lot has been written by several authors in English, Dutch, French, German, Japanese and Spanish about travel *A. fulica* has made since it left its original place (Mead, 1961). However, there is no record of any publication in this regard in any Indian language. A review of literature has revealed many conflicts and discrepancies mainly due to misinformation. Mead (1961) has clarified and explained in order to arrive at a better understanding.

Introduction, Dispersal/Distribution in India

According to historical records, W.H. Benson collected living specimens of *A. fulica* from Mauritius in February 1847. He carried these to Calcutta and released in Chowringhie Gardens the following April (Mead, 1961). After his return to England, he got the news that it had multiplied and was spreading in neighbouring states (Gude, 1914). Captain Hutton took some specimens from Calcutta to Mussoorie (Uttar Pradesh) at a height of 2211.6m (6,000ft), where they continued to breed but "a severe winter exterminated them (Gude, 1914). Within a decade of its introduction in Calcutta, reports of its spread started coming in. The first report was from Barrackpore (West Bengal), 25 km north of Calcutta and second was from Rajmahal (Bihar), 150 km

1. Release of *A. fulica,* after collecting from Mauritius, on the Indian soil, was a clear violation of quarantine laws, the consequences of which are being faced by the Indians.

north-west of Calcutta. Later, reports of its spreading in other parts of the country started pouring in (Raut, 1981).

This snail is now found in considerable numbers in the northern and eastern parts of West Bengal but rarely in the western districts. It is now found in western parts of Assam particularly in the Kamrup and Golpara districts. The author has himself observed this snail in large numbers in Guwahati. Behura (1955) reported this snail from Balasore district of Orissa and Singh and Birat (1969) from the districts of Darbhanga, Muzaffarpur, Motihari, Purnea, Saharsa and Santhal Parganas of Bihar. It is reported to have been introduced in Palghat (Kerala) (Mead, 1979). The author has observed the presence of this snail in Madras (Tamil Nadu), Calcutta, Barrackpore (West Bengal), Guwahati (Assam) and Baripada (Orissa).

On the Andaman and Nicobar group of islands, these were introduced by the Japanese for eating during the Second World War, during their occupation of these islands. This was narrated to the author by many survivors of Second World War at Port Blair (Andamans). They clearly stated that this snail did not exist on these islands before Second World War. According to the observations of the author and his co-workers, in addition to Port Blair and villages of the main island of South Andaman, this snail is now found on Long Island, Rangat, Mayabandar, Diglipur, Car Nicobar, Camorta, Nancowry and Great Nicobar. In the absence of proper domestic quarantine measures, this snail must have reached some more inhabited islands also with the transfer of seedlings and saplings from one island to another.

Raut and Ghose (1984), carried out extensive survey of the country in relation to dispersal/distribution of *A. fulica* and have presented the latest position of this snail in their publication of 1984 (Raut and Ghose, 1984).

As a result of their survey, Raut and Ghose (1984) have stated that in West Bengal it has spread from Calcutta to Balurghat, Gour, Jangipur, Alipore Duar, Bangan, Bahrampur, Bamungola, Bankura town, Bishnupur, Bolpur, Chakda, Contai, Dinhata, Egra, Golpakher, Harish Chandrapur, Hili, Jadavpore, Jalpaiguri, Kakdwip, Katoa, Kumargunge, Mathabhanga, Midnapur, Raigunge, Ramnagar, Saithia, Tapan, Ballygunge, and Coochbihar; the population varies from $8.12/m^2$ in Balurghat to $32.68/m^2$ in Coochbihar. In the North-eastern region snail infested pockets have been studied in 9 locations namely at Silonijan, East Sarani, Golaghat, Jhalukbari, Lachitnagar, Panbazar,

Tezpur, Ulubari and Jorhat in Assam. The population ranged from 8/m^2 at Silonijan to 26/m^2 at Jorhat. In Nagaland, snail infested pockets have been located at two places i.e., at Chumukedima (4.12/m^2) and Dimapur (24/m^2). In Meghalaya, at Nongpoh, in Tripura at Kumarghat, in Manipur at Imphal snail infested pockets were located with the average populations of 11, 17.22, 13/m^2 respectively.

Mead (1979) states that there are reports of its presence in Nagaland, 'but this still needs confirmation'. Now Raut and Ghose (1984) have confirmed its presence in Nagaland.

Besides West Bengal, Bihar is the only other state having snail infested pockets at a large number of places. Such pockets are spread over 32 places namely Telecocolony, Bhagalpur, Begusarai, Binda, Chakradharpur, Chapra, Darbhanga, Deoghar, Dhanbad, Dhruba, Dumka, Gumla, Jamshedpur, Jogbani, Jainagar, Katihar, Madhubani, Monghyr, Motihari, Muzaffarpur, Parsudihi, Pakur, Purnia, Pusa, Rajnagar, Saharsa, Samastipur, Simdanga, Sitamarhi, Vaishali and Chaibasa, the average population ranging from 6/m^2 at Telecocolony to 38/m^2 at Chaibasa.

Damage to crops by *Achatina* in Orissa was reported as early as in 1955. Now it is reported from Anandpur, Baliapal, Balasore and Baripada, the average population being 22.32, 25, 27 and 32/m^2 respectively. Raut (1986) reported the presence of *A. fulica* in most of the horticultural gardens of Cuttack, Bhubaneshwar and Konark (Puri).

Report of the presence of this snail in Uttar Pradesh appears to be very recent. Up to 1979, we had no report of the occurrence of this snail neither from the Department of Agriculture, Uttar Pradesh nor from the Plant Protection Department. Now Raut and Ghose have reported this snail from Bijnor, Dhampur and Moradabad, the average population recorded being 19.22, 27.32 and 34.40/m^2 respectively.

In the southern region of the country there were already reports of *A. fulica* from Kerala and Tamil Nadu. Raut and Ghose have recorded this snail from Karnataka also. Thus according to their records, the snail has been found in Malappuram, Palghat and Calicut in Kerala, the average population being 15, 24 and 34/m^2 respectively. It has been recorded from Madras, Annamalai Nagar, Pollachi and Tambaram, the average population being 15.12, 22.11, 22.44 and 34/m^2 respectively. As regards Karnataka there is record of snail infested pocket from Bangalore only, the average population being 9/m^2. There are no reports of this snail from Andhra Pradesh so far.

Thus the states/union territories not yet reported to be having giant

African snail-infested pockets are : Sikkim, Mizoram, Himachal Pradesh, Punjab, Haryana, Chandigarh, Delhi, Jammu & Kashmir, Rajasthan, Madhya Pradesh, Gujarat, Maharashtra, Goa, Daman & Diu, Dadra and Nagar Haveli, Pondicherry, Lakshadweep and Andhra Pradesh. Hilly tracts of Uttar Pradesh, Bihar and West Bengal also continue to remain free from this snail. A single record from Nagpur (Mead, 1979) has not been confirmed by Raut and Ghose (1984) nor the records of Kulkarni and Nagbhushnam (1973) from Marathwada region (Aurangabad) of Maharashtra state. However, the author is in a position to confirm the presence of *A. fulica* in Dapoli of Konkan region of Maharashtra as he received live specimens of this snail from Prof. R.B. Dumbre of Konkan Krishi Vidyapeeth, Dapoli, Distt. Ratnagiri on 14-9-1989.*

In India, the apathy shown towards this serious pest is deplorable. Neither any positive steps are being taken to check its further spread to other areas nor any regular research work is being done any where in the country-not even in the most heavily infested areas of Andamans. No adequate steps are being taken for its management even on the basis of already worked out methods of its management.

Dispersal/Distribution in other Countries

Bangladesh

Amongst the neighbours of India, its presence in some districts of Bangladesh (formerly East Pakistan) has been reported. In undivided India (till 1947), it was a part of Bengal where the snail was first introduced in the country; so its presence in Bangladesh is natural.

Myanmar (Burma)

Mead (1961) reports that this snail has reached Myanmar either from West Bengal or from Malaysia through the Japanese forces during their occupation.

Sri Lanka

According to Green (1910) a Conchologist named Oliver Collett

*According to a report published in *Deccan Herald* – of 14-3-91, this snail has been now located in Bombay also.

introduced the giant African snail on this island at Rozelle (near Watawala). He got the specimens either from India or Mauritius. Although as many specimens as possible were collected and destroyed, yet they are still in every province of Sri Lanka (Mead, 1955). However, there is now (1978) nothing like enormous infestation and since about 1970 "gradual spotty resurgence of these populations in certain areas" was observed (Mead, 1979). May be the Leucodermia–like disease might be spreading, leading to the population decline of the snail in this region.

Maldive Islands

In 1957, *Achatina fulica* were seen on Hitadu and Gan Islands in the Addu Atoll, but no snails were observed on Ile du Coin and Peros Banhos Atoll. The presence of the snail on some islands of this group is probably due to large amount of commerce between these islands and Sri Lanka (Mead, 1961).

China

They were first located in the Amoy University compound in June 1931 by Herklots. It is more likely that the snails have reached the mainland from the Amoy island (Mead, 1961). The snail is now found in great numbers almost every where in the open in Kwangtung and Fukien Provinces especially Pao-an in Kwangtung, Hainan Island and Tung-shan and Ho-nan outside Canton (Mead, 1979).

Hong Kong

The date of original infestation of *A. fulica* in Happy Valley-Sookunpoo area in Hong Kong appears to be 1937 (Jarrett, 1949). Jarrett believed that the snail had been brought from the Amoy island of China by the Chinese duck-keepers. Within a decade the infestation has spread to a number of places (Dean, 1950). Now it is a 'major pest to local farmers and gardners'. However, because of the unusually dry and cold weather prevailing for about twenty years according to the available records, the population was drastically reduced in spring 1976. Since then the incidence of the snail continues to be low (Mead, 1979).

Taiwan (Formosa)

This snail was introduced into Taiwan by a Taiwanese Government Official Kumaichi Shimojo in January 1932, but all the specimens died mainly because of the cold weather. In April 1933 more specimens were brought in and those which survived and bred, later served for introduction into Japan and Micronesia (Esaki and Takahashi, 1942). By 1940 the snail had spread almost throughout Taiwan (Kaburaki, 1940). Now it is reported to be "abundant in agricultural farm, vegetable garden and citrus grove near the hill sides during early May to late November" (Mead, 1979). Its utility for this island will be discussed elsewhere in this book.

Thailand

Boettger (1951) is of the view that this snail entered Thailand from Malaya (Malaysia) 'at least by 1937'. They are reported to be confined to the peninsular strip of the country. Here again, as in the case of Hong Kong, Chinese duck-keepers were responsible for its introduction, this time from Malaysia as ducks feed on the snail. In December 1957, Mead found heavy infestation in Bangkok (Mead, 1958 a, b). In 1964, this snails was seen near the town of Chiang Mai. Now the snails are found in the town and also about 5 km away in the Chiang Mai University Campus. It continues to spread around Chiang Mai. It has also spread across the river Mae Ping from West to East (Mead, 1979).

Cambodia

Brumpt *et al.,* (1968) have reported that this snail is found in the gardens of Phnom Penh.

Vietnam

Boettger (1951) and Rees (1951) both are of the view that *A. fulica* must have reached Vietnam by 1937, whereas Fischer and Huynh (1966) are of the view that it has been in Vietnam since 1935.

Malaya (Malaysia)

South (1923, 1926) and Jarrett (1923, 1949) are of the view that this snail first entered this country in 1911 in Kedah, either from India or Ceylon (now Sri Lanka) or may have come from India, via Burma. From Kedah in 1922 the Chinese duck-keepers took the snails to the Province Wellesley from where it spread quickly to the South in many areas of the peninsula. Now it is very common in the South-eastern part of the state of Perak. It is reported to be causing heavy damage to the kitchen gardens. They have also been reported from Pahang (Lim and Onar-Ahmad, 1969) and Kuala Lumpur (Lim *et al.*, 1965).

Singapore

Jarrett (1949) is of the view that the giant African snail could not have been introduced into the island before 1917 i.e., six years before its introduction into Northern part of Malaya. Mead (1961) is of the view that there is the highest likelihood that Singapore received this snail from Malaya. According to the information received by Mead (1979) this snail continues to be a pest of vegetables, ornamentals and orchids.

Indonesia

This is an island nation and comprises of a number of islands, the comparatively bigger islands being Sumatra, Java and Borneo.

Available records give conflicting reports with regard to the introduction of this snail into Java. According to Kalshoven (1950), this snail was first brought into Bogor from Singapore in 1922, whereas according to another report they arrived from Ceylon (Sri Lanka) in 1925 along with grass plants. Leefmans (1933 a, b) is of the view that these were introduced into Sukabumi a suburb of Batavia (Jakarta) in 1930-31 and were taken into Batavia by plant breeders in 1933. Butot (1952) stated that the whole of Java is infested except the mountainous region beyond 1000m height and some uncultivated parts below that level.

As regards Sumatra this snail is reported to have appeared first in north-eastern Sumatra in 1930-31. It is however, speculated that the snail had been on this island since 1921. In South-eastern region it was seen in 1931 or 1932. (Meer Mohr, 1935, 1940, 1948; Weel, 1948 a).

In East Borneo it was found in Balikpapan in 1921 (Latif, 1933). Later it was reported from other parts of the island also. As stated by Mead (1979) *Achatina* was not considered to be a serious pest, according to the information available to him up to 1972. It is reportedly found 'everywhere in Indonesia and common in West New Guinea, Biak island, Ternate, Halmahera, South Celebes, East Kalimantan, Southern Sumatra, Java and Bali (W. Soegeng Reksodihardjo quoted by Mead, 1979). It has been considered the "most important pest in vegetable growing". The snails are also used by duck-keepers to feed their ducks.

The Philippines

Pangga (1949) sets the date of entry of this snail into these islands as 1942. Here again the Japanese soldiers are said to be responsible for introducing it into these islands from Formosa (Taiwan). By 1949 most of the provinces of Luzon were infested including Batangas, Bulacan, Camarines Sur, Cavite, Laguna, Nueva Ecija, Nueva Vizcaya, Pampanga and Rizal (Mead, 1961). Choy (1963) has reported damage in the Makiling National Park to 35 species of plants and 7 native tree species. Reforestation work and nursery operations are hampered by these snails "because they destroy seedlings before they can attain tree size" (Mead, 1979).

Sarawak States (Malaysia)

According to Jarrett (1931, 1932) this snail was introduced in Kuching district of Western Sarawak in 1928, here again, by the Chinese interested in poultry farming. Tom Harrisson (1952, quoted by Mead, 1961) has stated that in the past twenty years these have spread steadily. Even as late as 1970, the snail population in Kuching was variable and often localised. Detailed studies of this snail in Sarawak have not been done; however, on pepper vines 10-20 snails per vine have been observed (G.J. Turner, quoted by Mead, 1979) and assuming 700 vines per acre (0.4 ha), the snail population comes to 7000-14000 per acre.

Japan

The giant African snail was imported from Formosa (Taiwan) into Japan many times. According to Esaki and Takahashi (1942) it started in 1935 but Kuroda (1949, quoted by Mead, 1961) considers the date as "previous to 1933" i.e., close to the time when the snails were first seen in Formosa. A number of articles published advocated the rearing of the snail for food and medicinal purposes and this encouraged large scale import of snails into Kobe, Osaka, Nagoya and Nara. However, in May 1936 Ministry of Agriculture and Forestry recognised this snail as a menace. Regulations were enacted immediately to confiscate and destroy all live specimens in Japan and prohibit the entry of further shipments. Thus the establishment of this pest in Japan was prevented and it is believed not to be present anywhere in Japan (Mead, 1961).

Mauritius Islands

According to Bequaert (1950) there are reasons to believe that *A. fulica* did not become established on Mauritius before 1800. According to a report of 1803 it is said that the wife of a Governor of the island ailing of the chest, had, on the advice of her doctor obtained snails of this species from Madagascar. She died shortly after but the snails spread over the island and they were even now very common. According to Benson (1858) a little before 1847, another species of *Achatina, A. panthera* was introduced into Mauritius by David Barclay. According to certain reports the new species has established itself over the lowlands up to 1200 feet above sea level and *Achatina fulica* has maintained its population at 1200 to 2000 feet above sea level (Mead, 1961).

As reported by Mead (1979) it is a 'serious pest' abundant especially in the humid localities of Central Plateau but during December to April, the rainy season of the island, it spreads to low lying regions also. They are common in the vegetable and flower gardens and damage the tree seedlings also in the nurseries.

Reunion

Bequaert (1950) is of the view that this snail reached Reunion early in the nineteenth century shortly after it became established on Mauritius (on Mauritius it is reported to have reached about 1803). Ferussac found it present on the Reunion in 1921. While Lesson (1830) feels that the snails reached Reunion from Madagascar, Bequaert (1950) feels that it came from Mauritius. The snail continues to be pest particularly in vanilla plantations. However, as a result of biological control through predatory snails the giant African snail population has "greatly decreased" by 1978 (Mead, 1979).

Seychelles

Dufo (1840) reported this snails on these islands in 1840. Dupont

(1935) reported these as being present on Mahe, Praslin and Silhouette. *A. fulica* has been on these islands for many years and is quite common. It is a serious pest of vegetable gardens on most of the islands. However, some of the more remote islands have remained free from this pest as the people keep a check on any vegetable material entering the islands (Mead, 1979).

Australia

Despite excellent quarantine regulations of long standing in Australia, *A. fulica* has entered Australia though as late as April 1977 when some students who were visiting Gordonvale situated at the south-east base of Cape York peninsula on the north-east coast of Queensland. It was identified as *Achatina fulica* (Coleman, 1977). Immediately an eradication campaign by locating and destroying was started. Between April to December 1977, 296 snails were officially recorded. This figure excludes eggs and specimen with less than 10 mm shell length. Its origin still remains unknown although heavily infested regions of Papua are only 800 km to the north. *A. fulica* continues to be intercepted in Australian ports.

Continental North America

In spite of the fact that *A. fulica* has already established on the Hawaiian islands, this pest has not yet established itself in the North American continent. This is a clear cut example of efficacy of the U.S. Quarantine but yet there are individuals determined to circumvent these (Mead, 1979).

The first major incident came close to starting an infestation in Mesa, Arizona (Mead, 1959b) but these were intercepted and destroyed. The second incident occurred in 1963 in Vancouver, BC., a third in Canton, Ohio in 1969. All these were intercepted and destroyed. The fourth incident occurred in 1966 when a young boy brought a number of live specimens, did not declare and took them to his grandmother in Miami in the peninsular Florida. The three specimens were released by his grandmother in her garden in June 1966 but it was only in 1969, more than three years after their release, that the resultant sizeable infestation was brought to the notice of the State Department of Agriculture. Later one after another four more, heavily infested foci were located up to July 1972 and destroyed. Finally in September 1972 a snail having 178 mm shell length and weighing 340g was located, 12 km west of the original infestation. Now, since no other specimen of *A. fulica* live or dead has been located since then "the area has been judged not to be an infestation site" (Mead, 1979).

Hawaiian Islands

In 1936 a young lady brought two specimens of *A. fulica* from Formosa in her baggage and released in her Oahu garden for "aesthetic reasons". Neither were these declared at the port nor discovered by the Port Inspector. The same year in November, some one imported specimens through mail from Japan in Makawao town in Maui for breeding and selling for medicinal purposes (Pemberton, 1938). The infestation were discovered only in June 1938 (Fullaway, 1939). On these islands, the snail quickly spread to other parts of the island and by 1944 there were 8 foci on Oahu and 6 on Maui. These were controlled for several years but by 1951 new foci made their appearance (Wong, 1951; Thistle, 1953; Weber, 1954). Infestation on Roundtop near Honolulu is most dense on record (Mead, 1959 b). Snails were found in possession of two men on Lanai but there are no reports of its infestation on that island (Pemberton *et al.,* 1939). In 1958, the snails were found well established near Mana (Kauai); two months later infestation was noticed near Hilo on Hawaii (Mead, 1958 a, b; Thistle, 1959).

Now, besides Oahu and Maui the other islands of the Hawaiian group reported to be infested with this snail are : islands of Hawaii, Kauai, Lanai and Molokai. According to Mead (1979) the overall picture is depressing and the snail has spread to all the major islands of this group.

In addition to the giant African snail infestations already referred to earlier in this chapter, there are a number of other places in the Indo-Pacific region, infested with this snail. Some of these are American Samoa, Guam, Comoro Islands, New Britain, New Ireland, New Caledonia, Papua New Guinea, New Hebrides, Ogasawara Gunto (Bonin Islands), Ryukyu Islands, Tahiti Islands, Tanzania, Wake Island and Trust Territory of the Pacific Islands.

3

FACTORS RESPONSIBLE FOR THE OCCURRENCE/SURVIVAL

There are a number of factors which play vital role in the occurrence/survival of land snails including *A. fulica.* These factors are interlinked and no single factor alone can be said to be responsible for the presence of the snail in any particular area. Some of the comparatively more important factors are discussed below:

Calcium

Van Cleave (1953) as a result of his studies on snails and their relation to soils, considers the presence of snails as an "indicator" of the availability of lime in the soil. Pilsbry and Bequaert (1927) and Oughton (1948) also consider relationship between the presence of snails and availability of calcium in the soil. These findings support the presence of large populations of *A. fulica* on the coralline islands of the Pacific. The soil of these islands is reported to be slightly alkaline having 7.2-8.7 as pH value (Mead, 1961). The Andaman and Nicobar group of islands also has a number of coral islands, which are infested with *A. fulica.* Atkins and Lebour (1923) have observed that the snails show preference for "slightly alkaline soils" (pH 7.0-8.0). However, the strongly acid soils of Ceylon (Sri Lanka) support equally large populations of *A. fulica* as the pH of soils in tea growing areas is about 4.5. Mead (1961) feels that calcium carbonate cannot last for long in such acidity especially when there is abundance of moisture. The calcium is locked up in silicates and in the form of iron and aluminium salts. The plants unlock this calcium as is evident from the fact that plant leaves in the acid soil area were found to contain as high as "40 per cent calcium in the ash". The snails were found to "feed extensively" on fallen leaves and this explains their source of obtaining calcium "independent of the supply of lime in the soil". Moreover

Cacao is being grown more and more in Sri Lanka because of more demand in the world market; Cacao needs less acidic soil (pH range 6.0-7.0) and this pH is obtained by adding lime to soil; while making the soil more favourable for cacao, it also makes it more favourable for *Achatina*. Burch (1955) recorded an increase in the number of land snails with the increase of calcium, magnesium and potassium.

Schmidt (1955) has "experimentally shown" that calcium carbonate is essential for the development of gastropod fauna. Besides being necessary for shell formation of the snail, calcium is also required for egg shell formation.

The shells of dead snails are added to the soil and the calcium cycle goes on. Raut and Ghose (1984), however, are of the view that the impact of factors including pH and CaO, on the distribution of land snails, is insignificant. They are of the view that *A. fulica* can survive "equally well" both in acidic and alkaline soils though a "preference for soils with high percentage of calcium has been noted". This is in conformity with the findings of earlier workers that *A. fulica* is abundant in the acidic soils of tea gardens of Ceylon (Sri Lanka) as well as in the Coralline islands of the Pacific and in the coral islands of the Andaman and Nicobar group, whose soil pH was towards alkaline side.

By controlling the amount of available calcium carbonate in different soil types, Voelker (1959) demonstrated differences in size, shape, colour, weight and the rate of growth etc. of the shells of the developing *A. fulica*. Schreurs (1963) not only demonstrated the important role of calcium carbonate in the normal development of *A. fulica* but also highlighted the importance of certain physical properties of the soil, presence of sufficient organic material in the soil and availability of enough green parts of plants. Raut (1981) also has reported that alluvial soil with a high percentage of calcium is preferred by *A. fulica* and the population declines gradually in black and mixed red soils. Alluvial soil and to some extent, mixed black and red soils can conserve moisture well, leading to luxuriant growth of vegetation, which, in turn, offers suitable humid environment for the snail. Raut and Ghose (1984) conclude that probably "nature of the soil does not influence the snail infestation directly". Actually, it is the capacity of the soil to hold water and thus influence temperature and humidity.

Temperature

Temperature is another factor which plays important role in the incidence of land snails including *A. fulica.* However, there is lot of tolerance to temperature variation, as is evident from its ability to undergo aestivation with the rise/fall of temperature. Chock and Nakao (1951) planned experiments in "standard refrigerators of various models". Nineteen out of 20 dormant snails exposed to 40°F (4.4°C) for 24 hours remained alive after 5 days; 7 out of 10 active snails exposed to a temperature of 30°F (–1.1°C) for 8 hours remained alive after 5 days and one out of 10 dormant snails, exposed to 20°F (–6.7°C) for 8 hours remained alive after 5 days. Mead (1961), however, states that in strong contrast to these survival rates in the preliminary series, further experiments showed "consistently complete killing" and the authors have not given any explanation for this difference in the results. Populations of *A. fulica* are able to withstand 43° ± 2°F (6.1°C-7.2°C) in Florida but its chances of establishment there are not bright. Mead (1961) reports that in Chichi Jima also the temperature is as low as 45°F (7.2°C) but the fact is that *A. fulica* in these cooler areas not only survives but thrives.

Singh and Birat (1969) have recorded the activity of *A. fulica* in Bihar up to 8.8°C. Olson (unpublished) has given 25.7°C as the optimum temperature. He has given the maximum high temperature as 28.7°C and minimum low temperature as 9°C for activity and thus for feeding and growth. Olson (unpublished) concludes that the expected size of the snail is "directly related to the duration of optimum temperature". Raut and Ghose (1984) have stated that *A. fulica* does survive at the temperature range of 0°-45°C, but for population build up a temperature range of 22°-32°C is required. Areas where difference in the temperature of winter and summer months is less, i.e., sea shore areas, islands and atolls, are favourable for the activities of the snail.

Altitudinal upper limit for the survival of *A. fulica* is 350m in Hawaii, 400 m in the Philippines (Choy, 1963), 600 m in Mauritius and 1200 m in Sri Lanka (Mead, 1955, 1961).

In India, Raut (1983 a) has reported decrease in the population of *A. fulica* with the increase in altitude. Thus, per square metre of infested pockets there were 26, 20, 17, 13 and 11 snails at the altitudes of 31 m (Jorhat-Assam), 50 m (Guwahati-Assam), 595 m (Kumarghat-Tripura), 850 m (Imphal-Manipur) and 1166 m (Nongpoh-Meghalaya) respectively. In West Bengal, Kalimpong (1209 m) and Darjeeling (2000 m), Shillong (1500 m) of Meghalaya, Kohima (3000m) of Nagaland and Ootacamund (2249 m) of Tamil Nadu were not observed to be infested with *A. fulica*.

Above 1200 metres height, the giant African snail does not survive. Hutton had taken *A. fulica* from Calcutta and released at Mussoorie in Uttar Pradesh; (Benson, 1858) at 2011.68m but these were exterminated because of a severe winter (Gude, 1914). Raut and Ghose (1984) are of the view that *A. fulica* may not establish in the hilly tracts of eastern and northern India above 1500 m as sudden lowering of temperature in common in these areas.

Explanation for this variability rests with the fact that lower temperature accompanies higher altitudes. Hence for the capacity of the snail to become established in newer areas, temperature or the degree of fluctuation in temperature, is probably the "most crucial" environmental factor, although a field observer may interpret it in terms of altitude rather than environmental temperature (Mead, 1979).

Humidity

Humidity is one of the most important factors for the survival and biological activities of the snail. Raut and Ghose (1984) also are of the view that humidity is the prime need for "terrestrial animals" (including *Achatina*). It is regulated mainly by rainfall and also by temperature and velocity of wind. Annual rainfall of more than 1400 mm is favourable for the snails. Raut and Ghose (1984) state that the snail population is governed more by the number of rainy days in the year i.e., frequency of rainfall than total annual rainfall. However, Bhattacharjee (1980) is of the view that temperature and rainfall have "no direct effect on the contagious distribution pattern formation in *A. fulica*".

Raut and Ghose (1984) have observed that in nature the active period of land snails is restricted to monsoon only. Lange, as early as in 1950, stated that rainfall apparently influences the activities of the snails more than any other single factor and oviposition is correlated with periods of extensive rainfall. In Andamans also the activities of *A. fulica* start with the onset of monsoon by about mid-May and continue up to about December end. However, after September-October the frequency as well as the total monthly rainfall goes on decreasing gradually and consequently, activities of the snail also i.e., feeding, mating, oviposition, hatching keep on declining and comparatively lesser and lesser number come out of their hideouts after dusk. When the monsoon ceases to be active, the activity comes almost to a standstill as the snails undergo aestivation, except at places where shade and moisture continue to be available regularly. The author has seen in Andaman, in April snails moving about after dusk near wells which are in shade and water keeps on falling regularly as people take bath and draw water for their domestic use. During the monsoon, the author has observed snails sticking to the damp walls in shade in large numbers during day time and after dusk all come down and start their routine biological activities. While humidity is an essential requirement for the survival of land snails, "pools of free water" must be avoided to prevent "overhydration". "Even shallow water can cause entrapment" because efficient crawling is hindered; deeper water may even drown most land snails (Russel-Hunter, 1979).

Raut and Ghose (1984) have, in simulated habitat, found the range of 80-95 per cent humidity, suitable for optimum activity of snails.

Although humidity and temperature both appear to be generally of major significance for the activities of *A. fulica,* humidity appears to play a more important role than temperature. That is the reason why this snail is found in the inland areas like the plains of Bihar and Assam, where, in spite of regular winter season and consequent fall in temperature during the winters, humidity continues to be available during most parts of the year. Even for hibernating/aestivating, they need moist shady place.

Vegetation

With plenty of rainfall, there is increase in humidity and consequently increase in the activity of the snail and there is plenty of growth of vegetation. Increase in vegetation besides conserving moisture and providing shelter to the snail, also serves as food in plenty. It also helps in reducing temperature and maintaining humidity—conditions favourable for the snail. However, Boycott (1934) does not consider vegetation as an important factor in the survival of the snail. Wolda *et al.,* (1971) also are of the same view as Boycott (1934). Raut and Ghose (1984) do not agree with the views of these workers and the author endorses the views of Raut and Ghose (1984).

Food

Food is another factor necessary for the activity of the snail and is both quantitatively and qualitatively very flexible. *A. fulica* has a tremendous range of acceptable food. In addition to a huge list of preferred and other host plants, it draws most of its food supply from dead decomposed animal and plant tissues. The snail feeding on fallen leaves, debris, animal dung and dead snails is a common slight in gardens. Raut and Ghose (1984) also have referred frequently to the snail feeding on fallen leaves of plants in the garden. Olson (unpublished) states that "feeding of *A. fulica* on live plants comprises 25.2 per cent of the total, while 74.8 per cent of feeding was done on scavenging type of food. It feeds on crushed bodies of its own kind, dead animals, decaying and rotting vegetation, dung, garbage, wet paper, cardboard and white-wash etc. More than 98 per cent of the food fed upon contained moisture.

However, this snail has immense capacity to survive without food for long periods. Lange (1950) observed that a snail (*A. fulica*) placed in a cardboard mailing tube lived for five months without food, except for some cardboard lining. Another snail lived for 5.5 months in an outdoor cage without food. Mead (1961) also states that it can go without food and water for months together. During hibernation/aestivation, they remain without food for many months. If the hibernation/aestivation period is prolonged because of prolonged drought and delay in monsoon, many do die but quite a good number survive also, in spite of the dehydration, they undergo during their inactive period during the drought/fall in temperature.

Light

Like all other land snails, *A. fulica* also is nocturnal and remains active during twilight also. Due to the nocturnal habit of land snails including *A. fulica,* they are able to take advantage of "cooler night temperatures and the presence of free water on plant surfaces in the form of dew" (Russel-Hunter, 1979). During day time also when the sky is overcast or it is drizzling and there is comparatively little light, then also the snails can be seen actively moving about. Because of its nocturnal habit it is able to avoid a number of "basically unavoidable environmental factors operative under normal diurnal conditions" (Mead, 1961). Sunlight is one such factor. Mead (1961) reports that the snail "dies quickly" when exposed to "direct sunlight". During the aestivating period, at Port Blair, the author activated a few snails, placed them in the open field without vegetation and exposed to direct sunlight around 11 a.m. All of them tried to crawl away but were not allowed to bore in the soil; while crawling all died within 30 minutes. Once a snail returning to its hideout a little late at dawn in November was interrupted by the author at Port Blair by withholding its further movement by about 20 minutes. When freed after 20 minutes hold up, it started moving and stopped on a wooden pillar of the portico of the Guest House. The pillar was then under shade but as the sun's rays fell on the snail, it moved further; it went on moving further and further up to about 9 a.m. when it reached a spot not exposed to direct sun's rays.

Raut (1983c) also states that the snail avoids direct sunlight and the snails resting in sites occasionally exposed to sunlight slowly shift their position with change in the direction of light. In other words *A. fulica* and other land snails can be said to be negatively phototropic. This, actually, is an adaptation to conserve water by reducing evaporation of water during day hours.

To sum up, Karlin (1961) has, with regard to land snails, very appropriately concluded that the interplay of factors like fluctuating temperature, availability of food, air movements, humidity, free moisture, incident light, age and size determines activity of the snail in nature "under conditions which cannot be completely duplicated in any laboratory experiment".

4

BIOLOGY AND BEHAVIOUR

General Characters

The giant African snail is a land pulmonate with a protective unichambered calcareous shell. The biggest-sized shell reported so far is 19.5 cm (Peterson, 1957b). The same author has recorded a snail with shell and fully stretched body measuring about 30.48 cm (12″). Srivastava and Abbas (1983) have recorded from South Andaman an *Achatina* shell measuring about 17.78 cm — the biggest size hitherto recorded from the Indian region.

The snail has an anterior conspicuous, fleshy head prolonged into a contractile snout. The Head bears two pairs of tentacles. Both the pairs are olfactory and tactile, but eyes are borne only at the tip of the longer, posterior pair. If the tentacles are lost by any chance, both the tentacles can regenerate — the one along with the eyes also. The head is joined directly to a ventrally-situated large flat, muscular foot.

Habit and Habitat

A. fuliça, like other land snails, is nocturnal and twilight lover. As a matter of fact the evening twilight is a signal for the snails to start their activities and the morning twilight is a signal for the cessation of their activities till evening. In short, generally their activity is restricted from dusk to dawn. During the day-light hours also it has been seen actively moving about when the weather is cloudy, the sky is overcast or it is drizzling. In any case, they always avoid direct sunlight. This snail is herbivorous and feeds upon a large number of plant species, but at the same time it is scavenger also because it feeds upon crushed bodies of its own species, dead small animals, decaying and rotting vegetation, dung, garbage, soil, damp paper and cardboard and white-wash (Olson, 1973). It has been observed rasping on shells of dead snails, egg shells and feeding on yellowed fallen leaves as these contain a high percentage of calcium. Ashes of dead leaves of tea growing areas in Ceylon (Sri Lanka) have been found to contain up to 40 per cent calcium (Mead, 1961). Olson (1973) calls *Achatina* an opportunistic, omnivorous and carpophagous feeder. He considers this snail to be basically a scavenger as 74.8 per cent of its food is scavenger-type and live plants form only 1/4 of their food (Fig. 4.1). More than 98 per cent of the food fed upon contains moisture. This snail inhabits damp habitats. In the rainy season, during day time, it is a sight to see these sticking to damp walls in shady places.

During the season of activity, during day time, they hide in a retracted state under fallen leaves, rotting logs, stones and in the holes of tree trunks. These have been observed resting on damp walls in shady places and also at the top of papaya and banana plants, on coconut tree-trunks and on the top of arecanut trees even at the height of about 10m (Srivastava, 1976). However, majority of these snails hide themselves in loose soils in shady places, under shrubs and railings. Raut and Ghose (1984) feel that there is "some sort of advance signalling system" regarding sudden change of weather. They have observed that in the vast majority of cases, they move to protected sites before the outbreak of a storm or a heavy shower.

Factors Responsible for the Snail's Activity

For the activity of land snails, humidity and temperature both are necessary but humidity plays a major role as is evident from the fact that even for hibernation/aestivation, they locate damp shady places, thick bushes and thick hedges etc. They are unable to withstand dry weather and high temperature. Even in the presence of adequate humidity, they are unable to tolerate very low temperature. Under both the circumstances i.e., (i) dry weather and high temperature and (ii) adequate moisture and very low temperature, they undergo hibernation/aestivation — the states of inactivity.

The ideal conditions for its activity are good amount of rainfall, humidity, shade and optimum temperature (Gupta and Doharey, 1985).

When the aestivating snails resume activity on the onset of monsoon, they start feeding on the food plants of their choice. Those which are sexually mature, start mating after locating a suitable partner.

FIG. 4.1: *A. fulica* in its natural habitat in South Andaman

Reproduction and Development

Sexual Maturity

Green (1911) states that *A. fulica* becomes mature in one year, when it is only half-grown. Leefmans (1933 c) mentions the age of six months for sexual maturity when calcium carbonate is available otherwise it takes nine months. Dun (1949, quoted by Mead, 1961) reported that snails raised by him laid their first batch of eggs at the age of 5-8 months. Weel (1948c) reports 7-8 months as the age of attaining sexual maturity. Garnadi (1951) has reported this period to be eight months. Daniel, B. Langford (quoted by Lange, 1950) states that on Guam, sexual maturity is attained in 147 days (about 5 months). Mead (1949) puts the time of sexual maturity as about a year. According to his report, male sexual organs mature earlier than the female organs. He recorded male sexual maturity before the snail is one year old, while the development of female sexual organs may take a few months more. Ghose (1959) found the snails attaining sexual maturity in the laboratory in six months but under field conditions it took more than a year because the growth gets almost suspended due to the intervening hibernation duration. However, it would also depend upon the time when the cold, dry weather approaches. Kondo (1964) states that the sexual maturity is reached at the end of the first year of snail's life. His smallest sexually mature specimen was 59 × 27 mm. Lange (1950) gives the shell size of sexually mature snails as 57-88 mm and 7-7.5 whorls. Meer Mohr (1949) gives shell-length as 80 mm. Schreurs (1963) found in West New Guinea that 'under favourable conditions *A. fulica* could go from "egg to egg in $6^1/_2$ months". Ghose (1959) gives the shell-size of sexually mature snail as 59.0 mm, but Raut and Ghose (1984) give the shell-size of the youngest sexually mature snail as 56.4 mm.

Mating

When the snails become sexually mature, they start moving about to look for a suitable mate. Raut and Ghose (1984) have described mating in *A. fulica* in detail. Mating takes place generally between snails of the same age group i.e., between snails of more or less the same shell-size. When they come across a suitable mate they touch each other with their tentacles, start moving together, and slowly come closer and closer to each other. The comparatively more active one then crawls over the other from the back side and secures position, the genital apertures of both come very close and are apposed. Then the intromittent organ of the climbing partner gets protruded out and it starts pricking with it the genital aperture of its partner. Within a few seconds the intromittent organ of the other partner also starts protruding out and making movements. Now the climbing partner inserts its intromittent organ into the vagina of its partner which in turn inserts its intromittent organ into the vagina of the climbing partner (Raut and Ghose, 1984). When the intromittent organs get properly fitted into the vagina of each other, they fall towards one side and copulation goes on (Fig. 4.2). During the process, mutual exchange of seminal fluid takes place and the sperms (Ghose, 1960) received by both the partners during copulation, are stored in the spermatheca for subsequent cross fertilization of the ova (Fig. 4.3). When copulation is over the intromittent organs are withdrawn from the vagina, the mated snails separate and move away.

FIG. 4.2: *Achatina* in copula

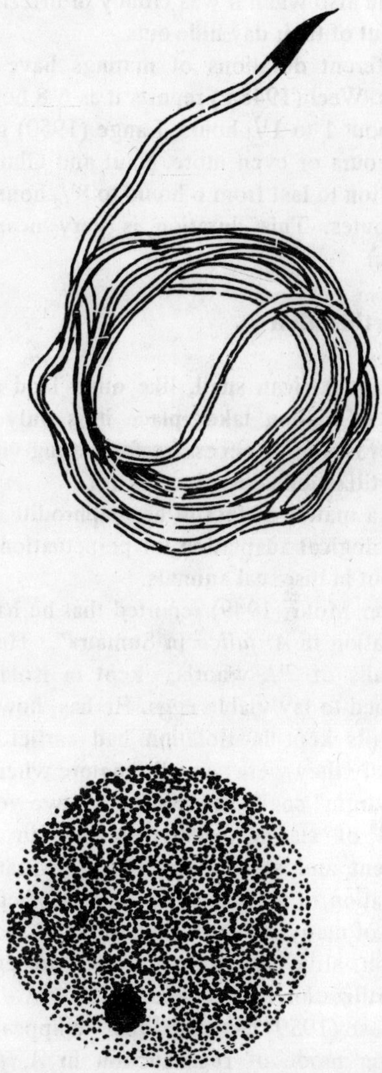

FIG. 4.3: Spermatozoa and ovum of *A. fulica*

Mating in *A. fulica* has been observed by the author on the Andamans during July-August, the peak season of activity of the snail. In some cases when the mating started around midnight they were seen in copula even in the morning. However, mating was observed during day time also when it was cloudy or drizzling due to which some snails come out of their day hide outs.

Different durations of matings have been reported by different authors. Weel (1948 c) reports it as 6-8 hours, Meer Mohr (1949) gives it as about 1 to $1^1/_2$ hours; Lange (1950) gives it as less than one hour to 24 hours or even more. Raut and Ghose (1984) have recorded the copulation to last from 6 hours to $8^1/_2$ hours, the average being 7 hours, 44 minutes. This duration is very near to that referred by Weel (1948 c).

Self-fertilization

The giant African snail, like other land snails, is hermaphrodite and whenever mating takes place it is only reciprocal. The question is whether mating is necessary for getting viable eggs or not and whether self-fertilization takes place or not.

As a matter of fact in hermaphrodite animals self-fertilization is a "physiological adaptation for perpetuation of species" and it cannot be ruled out in bisexual animals.

Meer Mohr (1949) reported that he had "repeatedly observed self-fertilization in *A. fulica* in Sumatra". His conclusions were based on the snails of $7^1/_2$ whorls, kept in isolation for 382 days and they continued to lay viable eggs. He has, however, not mentioned whether the snails kept in isolation had earlier mated or were virgins, for obviously they were sexually mature when they were isolated. Later he kept "virgin" snails in isolation for two years and none laid eggs. Then instead of stating self-fertilization in the snail he modified his statement and stated that a single, mating was "sufficient for the fertilization of a number of batches of eggs laid intermittently over a period of many months" (Mead, 1961). Leeuwen (1932) kept a snail in isolation alive for 9 years and no eggs were laid showing no self-fertilization having taken place.

Ghose (1959) has stated that "it appears that self-fertilization was a common mode of reproduction in *A. fulica*". He isolated 15 mm shell-sized snails; after attaining sexual maturity normal young snails hatched out from the eggs laid by the isolated snails. Since the snails were kept in isolation from 15 mm shell-size when sexual organs were hardly developed, mating is ruled out.

Kondo (1964) considers "self-fertilization as a proven point"; he states that copulation does occur but is apparently an unnecessary function in reproduction in *A. fulica.* Out of 50 young ones hatched out from the eggs he segregated 10, 4-day old infants. After two weeks, only 5 remained in the vials, the other five crawled out. Later these 5 were reared; finally only 3 remained in isolation and only one laid "10-15 small white infertile eggs". So he could not ascertain whether self-fertilization took place or not. Kekauoha (1966 a) kept in isolation 6 snails for four years and none laid eggs. Olson (1973) isolated 7 infants soon after hatching. These laid a total of 84 eggs; out of these 3 were found viable; out of these 3, one egg died before hatching. The infants of the other two died soon after hatching.

Raut and Ghose (1984), on the basis of their latest investigations have confirmed mating to be essential for reproduction of *A. fulica.* This is contrary to the earlier report of Ghose (1959) when he had stated self-fertilization in *A. fulica* to be a common mode of reproduction. Soon after hatching, they kept snails in isolation for 21 months and none laid eggs. As soon as they were allowed to mix and mate, all laid eggs. Thus besides Raut and Ghose (1984), Weel (1948 c), Kekauoha (1966 a) and Breckenridge and Fallil (1973) all are of the view that in spite of *A. fulica* being bisexual, mating is essential for reproduction.

Olson (1973) represents a very balanced view. He states that 'in spite of *A. fulica* being hermaphrodite "for all intents and purposes, cross fertilization is necessary for the laying of a sufficient quantity of eggs to ensure perpetuation of the species". He states that self-fertilization does occur but virgins lay less than 10 eggs at a time; most of the laid eggs are sterile and the possibility of a snail becoming sexually mature from a self-fertilized egg and reproduce is "quite low".

It appears that in spite of the fact that workers are of the view that mating is essential for reproduction, on certain occasions and under certain circumstances, self-fertilization also does take place, though it cannot be a common mode of reproduction.

Oviposition and Hatching

Raut and Ghose (1984) sate that eggs are laid generally in "isolated mass". Egg mass has to be covered with sufficient mucus to prevent desiccation. In isolated eggs, due to loss of water albumen becomes viscous and the developing embryo dies. Eggs are laid generally in loose soil a little under the soil surface. The snail digs a hole about 3-5 cm deep and 2.5 cm wide (Raut and Ghose, 1984). As soon as the hole is ready for oviposition the snail starts laying eggs "in heap" and a large amount of mucus is discharged which helps keeping the eggs together. Egg laying takes about 20 to 46 minutes. After egg laying is over, eggs are covered by the snail with loose soil.

Kekauoha (1966b) has observed that, before oviposition, the snail would burrow 1″–3″ (2.5-7.5 cm) deep into the soil and then lay eggs. Some times snails would oviposit their eggs only 1/4″–1/2″ (0.6–1.25cm) deep into the soil, and the main body of the eggs remains exposed. He observed that egg-laying took 1–4 days. After the ovipostion is complete, the ovipositing snail would rest from 3 hours to 3 days before resuming the normal activity.

The author has observed on the Andamans that although the general trend is to lay eggs a little under the loose moist soil surface, quite often in shade, but in a number of cases eggs are laid on moist soil surface and if they are laid on soil surface during the rainy season when it rains continuously, for days together, these eggs also hatch in the usual way. If however, eggs are laid on soil surface but not in shade and it it does not rain for a few days, the embryos within the egg shell die and do not hatch. During the rainy season, under field conditions, the eggs generally hatch in 7 to 10 days.

Mead (1949) has reported retention of eggs inside the uterus so that the eggs hatched within a few hours after oviposition. However, Lange (1950) did not observe it. Ghose (1963 a) states that eggs with embryos in different stages of development are laid hence the period of hatching varies; some eggs hatch even within a day after they are laid.

Lange (1950) has reported the interval between oviposition and hatching to be 6 to 12 days in the laboratory on Saipan, whereas in the field, it was about 11 days. Ghose (1963 a) has stated that in a brood, the first young snails came out of the eggs within 1 to 14 days after oviposition. He reports that the healthy ones come out first and weaker and smaller ones "especially from the small eggs emerge later". Kondo (1964) has reported this gap to be 1 to 15 days. Kekauoha (1966 a) has reported incubation period to be 1 to 17 days. Raut and Ghose (1984) also reported the gap between oviposition and hatching to be 1 to 17 days during their study of the snail with shell-sizes ranging from 56.4 to 104.9 mm (Tables 4.2 to 4.5).

Eggs — Number, Size and Shape

Mead (1961) states that as the age advances, there is probably decrease in productivity. Olson (1973) states that there is "decrease in fecundity and increase in sterility" with the lapse of time from sexual maturity.

Green (1911) reported 100 eggs in the first year and 500 in the second. He has not referred to the maximum number of eggs laid per snail. Meer Mohr's (1957) work also leads to this conclusion. Pilsbry (1919) has reported that 196 eggs are laid in loose soil just beneath the soil surface. Jarrett (1931) reported from China that in the first year 100, at the end of second year 200-300 eggs and they produced a total of 1000 eggs during the 3-year period. Meer Mohr (1949) has reported the number of eggs laid per brood as 82-315; Lange (1950) reports the number as 93-184 in the field and 62-243 in the laboratory. Rees (1950) refers to laying of 120 eggs in Kenya; Mead (1961) refers as 56-252 and Ghose (1959) as 27-356. Kondo (1964) mentions 27-356 eggs per clutch in the first year; the first clutch is the smallest, the clutch increases in number in the second and third years and declines in the 4th and 5th years. Total number of eggs laid in life time is 1000. Kekauoha (1966 b) has recorded 5-6 clutches of eggs per snail with an average of 213 eggs/clutch. The maximum total number of eggs laid per individual per season has been reported to be 1817 eggs. Olson's (1973) four of the mated snails laid 2848 eggs from January 5 to July 16, 1972. Raut and Ghose (1984) have referred to the laying of total number of eggs as 143-183, 354-568, 463-732 and 506-653 per year (Tables 4.2 to 4.5). Mead (1961) has reported the number of eggs laid per snail to be 300 in Micronesia, 200 in Ceylon and in Saipan the small-sized snails laid 30-60 eggs per snail. Atoda gave the maximum number of eggs laid per snail as 733 (reported by Hatai and Kato, 1943). This appears to be the total productivity of the snail with a shell-length of 66 mm. Mead (1961) reported that in the field "under optimum conditions", this species (*A. fulica*) will lay a batch of eggs "every few weeks" until the advent of unfavourable conditions.

The average number of eggs recorded per egg mass during different months of the year at Port Blair has been presented in Table 4.1.

Lange (1950) has observed on Saipan eggs to be nearly globular 4 × 5 mm in size. They are white when laid and later turn yellowish. Eggs on oviposition, are covered with mucus but within a few hours, the mucus dries up and the eggs become dull white (Lange, 1950). Kondo (1964) has mentioned the "average size" of the normal eggs to be 5.5 × 4 mm. About 2.5 per cent eggs are small-sized (3.5 × 3.0 mm). Kekauoha (1966 b) has reported the size of infertile eggs to be 2 × 2 to 4 × 4 mm and of fertile eggs to be 5 × 4 to 6.5 × 5.5 mm. Raut and Ghose (1984) have referred to "most of the eggs" as being ovoid, (Fig. 4.4) their size being 4.0-6.2 mm × 3.0 × 5.1 mm. They found 3.1 per cent eggs to be smaller in size and round in shape, size being 2.7 × 1.9 mm. Most of the eggs were deep yellow, the rest being light yellow to whitish.

Viability of the Eggs

Ghose (1963 a) has recorded viability of the eggs to be 82.5 per cent in those having embryos in the late stages of development and 26 per cent in those having embryos in the early stages of development.

Kekauoha (1966 b) has reported the viability of eggs to be from 0 to 100 per cent, the average being 91.1 per cent. He also reported high viability in the eggs with late stages of development.

Development

Development of different stages of *Achatina* has been studied by Ghose (1963 a). As a rule larval stages in land snails are completed within the egg and young snails resembling the parents hatch out. Rees (1951) has recorded at the time of hatching, cracking of the shell with a sharp click.

TABLE 4.1: **Average number of eggs recorded per egg mass at Port Blair from 1972 to 1976 in different parts of the year (After Gupta and Doharey, 1985).**

Month	1972	1973	1974	1975	1976	Average of 3/4/5 years
January	NR	NR	97.8	85.4	86.1	89.7
February	—	—	—	—	—	—
March	—	—	—	—	—	—
April	—	—	—	—	—	—
May	NR	121.0	105.1	95.2	87.9	102.3
June	NR	116.5	80.5	94.1	88.7	94.9
July	NR	112.4	94.9	95.7	97.2	100.1
August	138.7	113.1	96.7	97.2	101.2	109.4
September	117.1	106.1	97.1	93.2	102.9	103.3
October	89.7	106.1	88.1	94.5	101.6	96.0
November	90.4	81.6	83.1	92.7	90.5	87.6
December	86.0	103.7	78.8	90.5	75.5	86.9

NR = Not Recorded.

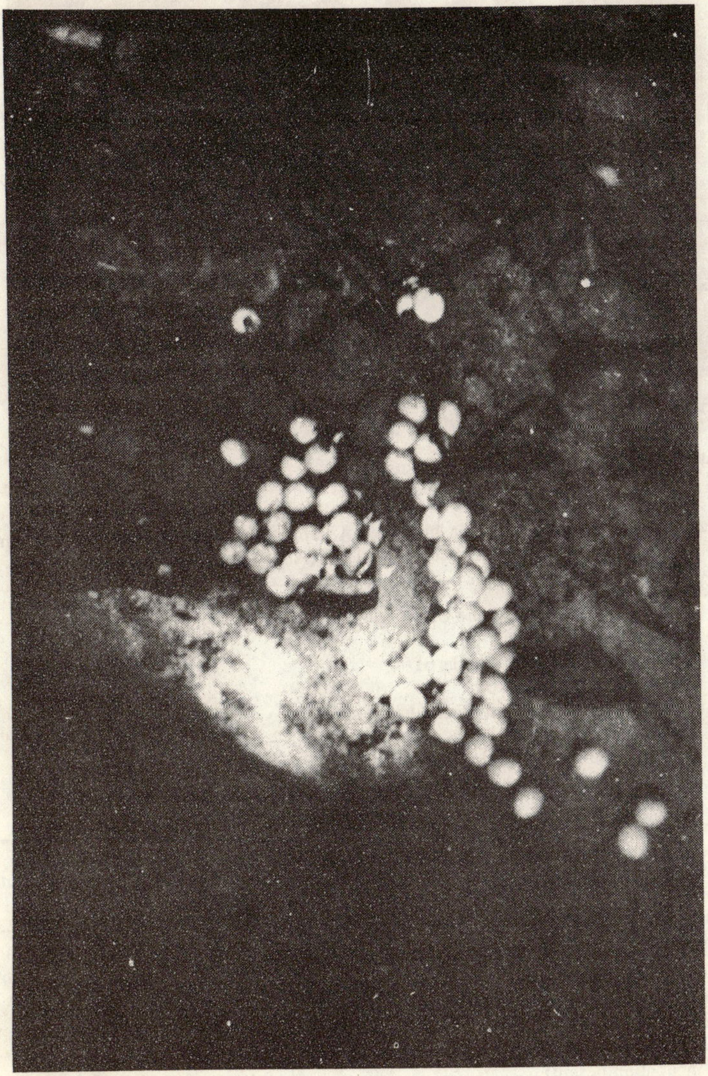

FIG. 4.4: (a) Eggs of the giant African snail

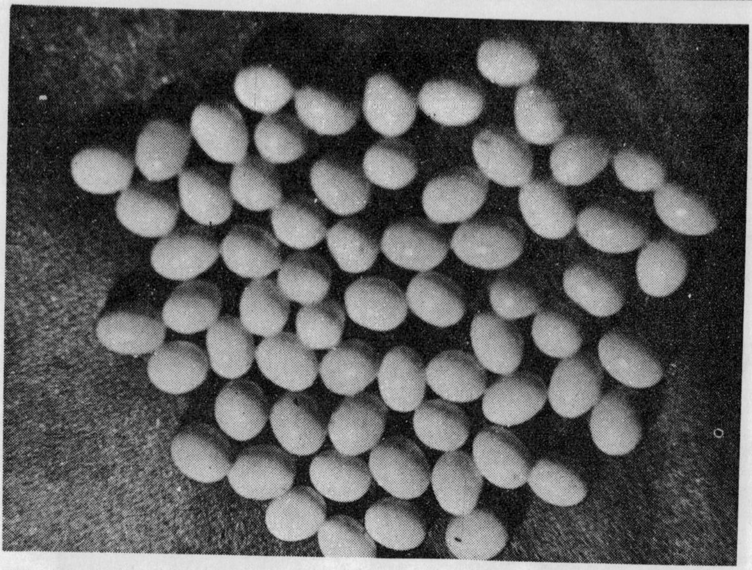

FIG. 4.4: (b) Eggs of the giant African snail

TABLE 4.2: Data on oviposition by mated *A. fulica* of shell length 56.4 to 66.2 mm during the breeding season of 1975 (After Raut and Ghose, 1984).

1. Mating during	June-July
2. Interval between mating and first oviposition (gestation period)	20 to 29 days
3. Oviposition during the months of	July-Sept.
4. Number of eggs per brood	46 to 120
5. Interval between oviposition and hatching (incubation period)	1 to 17 days
6. Total number of eggs laid per snail	143-183
7. Total number of young ones hatched from the eggs laid per snail	139-175

TABLE 4.3: Data on oviposition by mated *A. fulica* of shell length 75.2 to 80.1 mm during the breeding season of 1975 (After Raut and Ghose, 1984).

1. Mating during	June
2. Interval between mating and first oviposition (gestation period)	20 to 23 days
3. Oviposition during the months of	July-October
4. Number of eggs per brood	100-231
5. Interval between oviposition and hatching (incubation period)	1 to 17 days
6. Total number of eggs laid per snail	354-568
7. Total number of young ones hatched from the eggs laid per snail	332-540

TABLE 4.4: Data on oviposition by mated *A. fulica* of shell length 90.1 to 94.0 mm during the breeding season of 1975 (After Raut and Ghose, 1984).

1. Mating during	June
2. Interval between mating and first oviposition (gestation period)	20 to 41 days
3. Oviposition during the months of	July-October
4. Number of eggs per brood	167-272
5. Interval between oviposition and hatching (incubation period)	1 to 17 days
6. Total number of eggs laid per snail	463-732
7. Total number of young ones hatched from the eggs laid per snail	448-678

TABLE 4.5: Data on oviposition by mated *A. fulica* of shell length 100.1 to 104.9 mm during the breeding season of 1975 (After Raut and Ghose, 1984).

1. Mating during	June-July
2. Interval between mating and first oviposition (gestation period)	20 to 32 days
3. Oviposition during the month of	July-October
4. Number of eggs are brood	214-344
5. Interval between oviposition and hatching (incubation period)	1 to 16 days
6. Total number of eggs laid per snail	506-653
7. Total number of young ones hatched from the eggs laid per snail	488-620

The newly-hatched snail's shell is globose, very thin, almost transparent, glossy, without any decussation and of uniform light horny colour. The first whorl is very small, but the next one is quite large. The size of this infant snail is 4 mm (Ghose, 1963 a). Olson (1973), has recorded the shell size of infant *A. fulica* at hatching to be 5 × 4.5 mm and with $2^1/_2$ whorls.

The colour of all the organs except eyes (black), kidney (cream-white) and digestive gland (light brown) of newly-hatched infants is almost glass-like. In the larger snails (4 mm and above), which constitute about 10 per cent of the total, black pigment in the form of small rods and dots is scattered irregularly on the mantle over the pleuropedal visceral mass. Reticulation on the body is entirely absent. The heart, kidney, and pulmonary vein are distinctly visible through the shell.

The infants start feeding on soil and broken egg shells. Lange (1950) also has reported that young snails feed on the egg shells for 3-4 days. As expected they are more active during nights. Ghose (1963 a) has observed that the infants not allowed to come in contact with soil "did not thrive well". He feels that soil may be providing certain requirements of the infants in their early stages of development. From the 4th day onwards, very light brown pigment begins to be deposited on the dorsal surface of the visceral stalk.

The size of the 8-day old infant is 6.5 mm. The ground colour of the shell is uniformly light horny. Narrow, almost parallel, very faint light brown vertical stripes run from suture to columellar lip. Black pigmentation on the mantle increases; the digestive gland becomes brownish black and occupies the first two whorls; kidney is yellowish and the foot is cream-white. Light brown pigment appears on the visceral stalk, on both the pairs of tentacles, and on the dorsal surface of the foot. Pigmentation is heavier on the dorso-lateral side of the visceral stalk, posterior to ocular (posterior) tentacles and at the junction of the visceral stalk and foot.

The infants are very active especially during rains. Feeding takes place mainly during nights. Leaves are eaten from the margins. They can consume comparatively hard substances like germinating pea and gram seeds.

Size of 16-day old infant is 8 mm and it appears thicker and more opaque with a flexible and greyish horny lip. White patches in the form of small rods and dots placed at regular intervals appear on the apical visceral mass. The digestive gland occupies almost the whole of the first two and one fourth whorls. The colour of tentacles is like that of the visceral stalk. Pigmentation on the foot becomes light brown and reticulation appears on it.

Although the shells of 24-day old infants are no more globose yet they are not yet elongate. The infant of this stage is of 12 mm size; it is very much similar to the earlier stage except that pigmentation of the body is deeper and reticulation on the visceral stalk becomes more prominent.

Size of 32-day old young ones is 15 mm; it is now slightly elongate but is no longer glossy in appearance and is light roseate in colour with increased opacity. The vertical wrinkles on the body whorl are slightly bent at the sutures. The brown stripes on the body-whorl are deeper but, instead of running from suture to columellar lip, stop at the middle, being slightly deflected at the end. The white spots on the apical visceral mass and the black pigment on the pleuropedal visceral mass, assume elongated appearance and form small scattered stripes. Pigmentation on the rest of the body approaches a deep brown colour. The foot is light brown. Reticulation on the foot and visceral stalk is very prominent.

Size of 46-day old young ones is 18 mm. The shell is elongate-ovate, the apex forming a broad cone. The vertical wrinkles are cut by a few incised lines. The brown stripes are broad and deeper. The digestive gland shifts its position, and the first whorl of the shell becomes empty. The pigmentation on the apical and pleuropedal visceral mass is no more visible. Pigmentation and reticulation on the body and foot increase, and colour of the foot is now brown.

Sixty-day old snail is now 24 mm in size. The shell is still partially transparent. The shape and pigmentation of the shell nearly approach those of an adult. Similarly, the reticulation on the foot and visceral stalk is almost like that of adult. The colour of the exposed parts of the body is deep brown.

Seventy-four-day old snail is 31 mm in size. The pigmentation of the body and the shell is like that of the adult.

Eighty-eight-day old snail is 38 mm in size. It resembles an adult in all respects.

Lange (1950) noted that under field conditions there was 'high incidence of infertility' which can be attributed to desiccation or other environmental factors. However, as reported earlier, the author and his co-workers did not observe any such incidence under field conditions, when the eggs were laid in moist loose soil, under shade, during the rainy season.

Lange (1950) has observed the newly-hatched infants of the snail remaining in a group. Ghose (1963 a), however, has not reported this. Lange recorded observations for 156 days. Snails increased from 2.5 whorls at hatching to 6.5 whorls. Out of 20, he could successfully rear only 8 and a summary of the result is presented in Table 4.6.

TABLE 4.6: Results of the successful rearing of *A. fulica* (Lange, 1950).

Age	Average number of whorls	Average weight (gm)
At hatching	2.5	0.04
25-day old	4.0	0.22
60-day old	5.0	2.14
121-day old	5.5–6.0	6.37
156-day old	5.5–6.5	8.96

Lange (1950) has tabulated (Table 4.7) results of the observations of D.B. Langford (specimens deposited at the Hawaiian Sugar

Planters' Association Experiment Station, Honolulu).

Growth rate in *A. fulica* was studied by Kondo (1964) which is shown in Table 4.8. He has given the monthly observations of the growth of 5 specimens.

Kekauoha (1966b) has reported that in 535 days, the snails grew in length from 5.5 mm to 132.5 mm and in width from 4.5 mm to 55.7 mm.

TABLE 4.7: Number of whorls and length of specimens of different age groups (D.B. Langford, unpublished)

Age in weeks	Number of whorls	Length (mm)
1	3.00	5.00
2	3.50	8.00
3	4.00	12.00
4	5.50	26.00
5	6.50	34.00
6	6.50	40.00
7	7.50	63.00

Activity of the Snail

The snail was generally active during the rainy season from dusk to dawn. Their activity has been observed to be maximum from 12 midnight to 1 a.m. (Table 4.9).

Observations taken all the year round during 1973-74 on the seasonal variation of population at Port Blair indicated that the population recorded in a 25 m² plot was maximum during August and September, minimum during January, February and March and nil during April (Table 4.10).

Longevity

Leeuwen (1932) kept a specimen of *A. fulica* alive in captivity for nine years which is a record. Weel (1948 c) is of the view that Leeuwen's specimen represents an "exceptional case". Pelseneer (1935) states that Gibbons (quoted by Pelseneer) reported the life of *A. fulica* to be "at least" two and a half years. Mead (1961), on the basis of the longevity of other pulmonates feels that the average life span of *A. fulica* is 5-6 years, which is a "conservative estimate". However, where specimen are in areas "perennially offering optimum conditions for protracted

TABLE 4.8: Growth rate of *A. fulica* from March 1956 to May 1957 (After Kondo, 1964)

(Size in mm)

1956										1957				
March	April	May	June	July	August	Sept	Oct.	Nov.	Dec.	Jan.	Feb.	March	April	May
4	5	18	37	63	92	107	115	118	125	128	129	129	129	129
1	6	18	38	63	94	106	110	111	111	114	116	116	116	116
9	5	17	40	64	92	102	107	112	113	114	115	115	115	115
10	5	16	39	52	91	103	106	107	110	110	112	112	112	112
8	5	18	37	61	91	101	103	106	108	109	110	110	110	110

activity", the normal life span may possibly be shortened by a year or more. Olson (1973) concluded from the study of his field colonies that the life span is less than two years and probably "close to one year". Mead (1979) is of the view that since Olson's colonies were maintained in cages, "unnatural variables may have been operative and may have influenced his results". Mead (1979) reports that "of the 17 specimens, sent to him the oldest ones were found to be 2-2.5 years of age". Later in 1972 he came across a specimen, which upon examination, proved to be close to 5 years of age. Mead (1979) states that his earlier assumption that the normal life span of this species is about 6 years "now seems essentially confirmed at least for Hawaii".

TABLE 4.9: Activity of *A. fulica* from 6 pm to 5 am during the peak season of its activity at Port Blair (After Gupta and Doharey, 1985)

Time/hour	Average snail population in 25 m^2 plot			Average snail population
	July*	August*	September**	
6 pm	45.25	39.25	120.25	68.25
7 pm	75.00	75.37	192.00	114.12
8 pm	95.88	106.62	288.50	163.66
9 pm	118.13	129.25	332.25	193.21
10 pm	133.13	152.12	363.00	216.08
11 pm	151.50	161.50	375.25	229.41
12 Mid Night***	143.30	159.62	358.25	220.39
1 am***	130.13	151.62	294.00	191.91
2 am	100.87	129.63	230.25	153.58
3 am	66.75	101.25	172.50	113.50
4 am	43.13	63.62	110.25	72.33
5 am	23.38	32.50	57.00	37.62

* Av. data of 4 years (1972, 1973, 1974 and 1975)

** Av. data of 2 years (1973 and 1974)

*** Peak hours of activity of the snail

Behaviour

Nocturnal

As already stated earlier, the giant African snail is nocturnal and during twilight it is generally seen coming out of the hiding places or retreating to the hide outs. After midnight, they start returning back after devastating the crops, except those who start mating late and the process is not complete by about midnight. Such mating snails return to their hiding places a little late, generally during the morning twilight and rarely even after sunrise.

TABLE 4.10: Seasonal variation in the population of *A. fulica* during 1972-73 at Port Blair (After Gupta and Doharey, 1985)

Month	Snail population in 25 m^2 plot					Population (average) per week
	1st week	2nd week	3rd week	4th week	Total	
May	70	105	40	135	350	87.50
June	115	55	61	85	316	79.00
July	85	89	86	108	368	92.00
August	120	109	260	125	614	153.50
September	112	138	130	105	485	121.25
October	89	105	134	75	403	100.75
November	115	113	115	110	435	113.25
December	97	67	35	42	241	60.25
January	19	16	9	12	56	14.00
February	5	6	1	nil	11	3.00
March	1	3	nil	nil	4	1.00
April	nil	nil	nil	nil	nil	nil

Gregariousness

The snail has gregarious habit. Eggs are laid not singly but in a mass, the young ones hatch out in a mass and start feeding gregariously. When the snails come out of their hideouts in the evening they move together more or less as a group. During day-time they are seen resting on coconut and arecanut tree trunks and on papaya near about the fruits. This day-time rest also indicates their gregarious habit. They hibernate/aestivate also huddled together giving indication of their gregarious behaviour.

Their behaviour towards dusk and dawn cannot be explained adequately. Even if we presume that they are able to get an idea about dusk by their eyes, how to explain their knowledge about the approach of midnight and that it is time to return.

Feeding Behaviour

The snail is a voracious feeder. During feeding also they exhibit gregarious habit with the result that the crop or plant which they attack after dusk ceases to exist by the next morning and requires resowing/replanting. They feed not only on leaves but also on flowers, buds, tender shoots and fruits. When the young ones feed on leaves they feed by making small holes in the lamina. The author once saw at Port Blair that in a few plots where okra was transplanted in the evening, there was no trace of transplantation in the entire field by the next morning. Besides feeding on living plant material, it also feeds upon dead/decaying plant and animal tissues. Lange (1951) states that this snail is "primarily" a scavenger. Olson (unpublished) also states that about 75 per cent of the food of this snail consists of "scavenger type food".

Procuring Calcium

To obtain their calcium requirement these have been observed rasping shells of dead snails. Not only that, they have been observed feeding on yellowed fallen papaya and other leaves also. Such leaves are reported to contain high percentage of calcium. How does the snail come to know of the high percentage of calcium in fallen leaves has so far not been explained by any Malacologist.

Mating Behaviour

Although the snail is a hermaphrodite animal, it does exercise some sort of choice in the selection of the partner for mating. When these want to mate they start moving about in search of a suitable mate and when they come across one, mating starts; if the choice happens to be wrong, they separate and start afresh search for a mate. Details of mating have been discussed elsewhere in this chapter. The entire behaviour can be attributed mainly to instinct if there is any thing like instinct in this pulmonate gastropod.

Parental Care

After oviposition the snail moves away and does not come back to the spot where eggs were laid. The eggs hatch in course of time and the young ones take care of themselves. There is nothing like parental care in the snail.

Hibernation and Aestivation

A. fulica also, like other land snails, is a cold-blooded animal, hence to protect itself against unfavourable climatic conditions, it enters into a state of inactivity. This state of inactivity to tide over the unfavourable climatic conditions, is known as hibernation/aestivation. In India, in most of the regions, this period of inactivity occurs twice a year — once in the winter season and again during the summers. However, in the seashore regions and on the islands of Andaman and Nicobar group, where there is no shivering winter period in the year, the period of inactivity in the life of the snail occurs only once a year. On these islands the author has observed that when the rains are over, dry weather starts, moisture is reduced, the period of inactivity starts and continues till the onset of monsoon by about mid-May.

Hibernation/aestivation is only a device to tide over the unfavourable climatic conditions mainly dry weather and fall in temperature. The snails have to avoid dehydration. Hora and Rao (1927) are of the view that hibernation and aestivation are "outwardly very similar" but from the physiological point of view "they must be quite different". Raut and Ghose (1984) are of the view that in hibernation the body temperature goes down very much and the rate of metabolism, respiration, heart beat etc. become slow whereas aestivation is mainly because of the lowering down of atmospheric humidity. Brand (1931, 1932) stated that water content of the snail *Helix pomatia* is "actually lower" during hibernation than during the periods of "normal activity". Kamada (1933) agreed with Brand and reported that the blood of hibernating *H. pomatia* "was isotonic with 0.69 per cent NaCl whereas the blood of active snail was isotonic with 0.50 per cent NaCl." Based on these findings, Howes and Wells (1934a,b) stated that hibernation is a "physiological phenomenon" which requires "elaborate metabolic preparations" whereas aestivation is "comparatively simple physiological phenomenon". A direct correlation between aestivation and reduced moisture in the environment has been stated by a number of researches. The observations of Howes and Wells (1934a,b) enabled the workers to explain the presence of aestivating snails even under optimum conditions required for the activity of snails. These Malacologists came to the conclusion that in the life of a snail, there was a tendency for the phase of aestivation to alternate with that of activity, even under "more or less constant external conditions" with the result that "at any time" some of the snails would be in either phase. According to them, aestivation was the result of "low water content" which may be brought about either "by dryness of the environment" or "by the natural hydration cycle of the animal".

Mead (1961) states that the inhabitants of Chichi Jima report that during the winter, *A. fulica* are found 10–12.5 cm (4″-5″) below the soil surface. In Ani Jima and Chichi Jima of Bonin islands, there is a well established population of *A. fulica.* Kondo (1950 a) reports that during the past thirty years, the temperature in these areas (Ani Jima and Chichi Jima) continues to remain as low as 45°F (7°C); in spite of this, whether it is hibernation or aestivation in these cooler areas this snail survives.

Raut and Ghose (1984) have reported that during summers, this snail aestivates when the maximum temperature rises to 28-30°C at the humidity ranging from 80-82 per cent. These workers have also reported that "in spite of favourable humidity" during winters, at a temperature of 8.3°C, 58 per cent snails undergo hibernation. The author and his co-workers have observed in the laboratories of the Indian Agricultural Research Institute, New Delhi, that *A. fulica* starts hibernating when mercury touches 11°C; by the time, the temperature comes down to 5.5°C and RH goes below to 65 per cent, all undergo hibernation. Snails with bigger-sized shells, hibernate earlier than those with smaller shells (Srivastava *et al.,* 1987); thus the snails with 10 cm shell-size weighing between 102-120 gm were the first to hibernate followed by those with shell-size 3.5-4 cm weighing 7-9 gm. Another land snail, the common garden snail, *Macrochlamys indica* with shell-diameter 1.8 cm and weighing 1.4 gm hibernated even later than *A. fulica* having shell-length of 3.5–4 cm. Thus it can be generalised that land snails with bigger shells and weighing more, hibernate earlier than those with smaller shells and lesser weights. This holds good for individuals of the same species also. This gives a clear indication that there is some correlation between the weight of the snail and its hibernation.

Raut and Ghose (1984) have stated that in India, in most of the regions, twice a year the rainy and dry weathers alternate. Generally from October to December, there is cold and dry weather. Towards the end of December or early in January, there is generally a brief period of winter rains, thus there is a temporary break in hibernation. Whenever there is rainfall of even 50 mm the hibernation/aestivation of the snail is interrupted by a brief temporary period of resumed activity.

Raut and Ghose (1984) as a result of their continuous observations of several years have stated that the period of hibernation/aestivation of land snails (including *A. fulica*) varies in different parts of the country. They have stated that in southern part of West Bengal, Bihar, Orissa, Tripura, Maharashtra, Andhra Pradesh and eastern part of Assam this duration is of about 8 months; in northern part of West Bengal, western part of Assam, Nagaland, Tamil Nadu, Kerala, Karnataka and Uttar Pradesh, this period is of 4-6 months duration whereas in Meghalaya, Manipur, it is of only 2-3 months duration. The author and his co-workers have observed on the Andamans that heavy showers are over by September-October but the rainfall comes to an end by December or early January. From January to mid-May, a dry spell persists and consequently the snails continue to remain under aestivation except where there is shade and water continues to fall regularly from well or other sources to provide the snails adequate humidity for activity.

Dampness/humidity/moisture is integral part of the life of *A. fulica.* Even for hibernation/aestivation they locate damp loose soil under bushes, hedges, crevices in the mud walls, under bricks, stones, dead/decaying leaves or twigs. Mead (1961) has reported that although aestivating snails are often found under very superficial "protective conditions", these have also been found under big rocks, deep in the hollows of trees, in the centre of rotten logs and high on tree tops and tree forks. Raut and Ghose (1984) state that aestivation at a height of 4 metres also is "common". In Ceylon (Sri Lanka) it was observed that even during "a brief dry spell" during the rainy season, these remained in, inactive state on tree trunks, usually at the height of 0.91 – 2.74 m (3-9 feet) but sometimes even at a height of 9.18-10.60 m (30-35 feet) from the ground level. However, under prolonged adverse climatic conditions, the snails would crawl down the trees and aestivate in normal sites in/on the soil surface.

Before hibernating/aestivating, the snail retreats the muscular part of its body inside the shell (Fig. 4.5). To avoid dehydration, the shell opening is closed by a "thin, fragile, calcareous epiphragm" formed by the secretion of the snail itself. The epiphragm has a slit over the pneumostome for permitting reduced respiration. Raut and Ghose (1984) have observed that while lying in the state of inactivity, each snail formed 6-12 epiphragms. Before the formation of the new epiphragm, it throws off the older one. They have found only one epiphragm at a time in one snail. Snails in which, the inactivity was interrupted because of rain and feeding, the maximum number of epiphragms recorded by Raut and Ghose (1985) is 14.

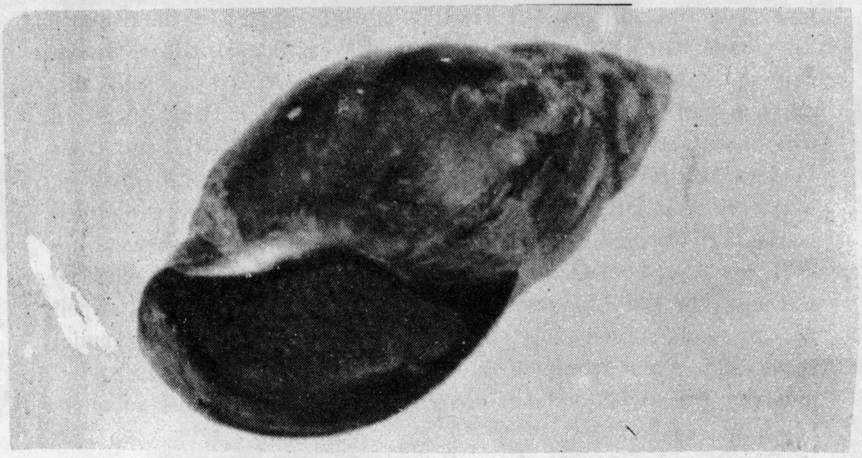

FIG. 4.5: The snail getting ready for hibernation — muscular parts of the body withdrawn within the shell before secreting epiphragm

Epiphragm Formation

Raut and Ghose (1984) have studied the process of epiphragm formation in *A. fulica* in detail.

In the secretion of epiphragm, foot and the mantle, both take part. The mantle-collar secretes maximum mucus, while the mantle and the foot secrete much less. Quite a good amount of mucus spreads "as a thin film" over the mantle and the foot and within two hours, a thin, more or less transparent membranous epiphragm is formed covering the organs but leaving a slit over the pneumostome. The epiphragm fits perfectly in the shell opening. The entire process takes about four hours.

Orientation of the Shell during Hibernation/Aestivation

While lying in the state of inactivity, orientation of the shell is very important. *Achatina* remains in inactive state within its shell with the shell opening directed downwards (Raut and Ghose, 1984). Closure of the shell-opening with a completely fitting epiphragm and lying with the shell-opening directed downwards helps the snail to avoid dehydration and maintain its body moisture at a level essential for remaining alive.

During aestivation/hibernation, the snails congregate and lie one over the other. Thus in a small space, more snails in the inactive state are accommodated. Raut and Ghose (1984) have reported that in an area of 135 m^2 in West Bengal, they could find 4-9 large aestivating pockets in which 17 to 61 hibernating snails were found. The number of small pockets was 18 and 5 to 14 snails were found resting there. Aestivation of 5 to 19 snails per m^2 area is quite common. Whether the pockets be big or small, the space available in the pockets was fully utilized and more snails could not be accommodated. Raut and Ghose (1984) have made similar observations in Tamil Nadu and Kerala.

Abbas and Gautam (1975) studied the aestivating population of the giant African snail in 1973 in the aestivating pockets of 11 villages of South Andaman. The work of studying the aestivating pockets in all these villages was continued in 1974 and 1975 also. Later a comparative account was published in 1978 (Gupta *et al.*, 1978). The highest population i.e., 102.39m^2 population was noticed in Makkapahar in 1973 (Table 4.11). Although the population went on declining from 1973 to 1975 in all the villages but Makkapahar and Calicut continued to stand first and second respectively during the three years with regard to the aestivating snail population (Table 4.11). It was also observed that the villages farther away from Port Blair town had higher aestivating snail population as is evident from the fact that Makkapahar and Calicut which are respectively about 20 and 14 km away from Port Blair, had very high aestivating population while villages like Dolliganj, Nayagaon, Shadipur, Schoolline and Pahargaon which are only 5 to 8 km away from Port, had much lower snail population.

Hibernation/aestivation is only a device to tide over the unfavourable circumstances but it is no guarantee against death as all the resting snails in the aestivating/hibernating pockets do not resume activity on the onset of monsoon. Quite a good numbers die because the minimum quantity of water essential for life does not remain in these because of over dehydration. If the monsoon is delayed and the drought prolonged, the resting snail population gets further reduced.

Resumption of Activity

The aestivating/hibernating snail population is of great significance as it is this population which is responsible for "restocking an infested area after the effects of normal or man-made adverse conditions" have disappeared (Mead, 1961). With the onset of monsoon, the inactive snails which have survived the stress and strain of adverse climatic conditions during hibernation/aestivation, become active again and resume their biological activities. Raut and Ghose (1984) state that more than 50 mm rainfall can terminate aestivation at any time. It takes 1-3 hours to resume activity. While resuming the activity, the epiphragm is pushed out by the foot.

TABLE 4.11: A comparative study of the aestivating population of *A. fulica* in South Andaman during 1973, 1974 and 1975 (After Gupta *et al.*, 1978)

Sl. No.	Locality		Years 1973	1974	1975
1	2		3	4	5
1.	Schoolline	a.	34.00	50.00	143
		b.	92.64	74.13	200.10
		c.	2429	1682	3672
		d.	26.21	22.68	18.35
2.	Pahargaon	a.	68	67	226
		b.	120.53	104.49	233.17
		c.	3810	2898	6202
		d.	31.61	27.73	26.59
3.	Austinabad	a.	62	59	207
		b.	106.76	54.39	268.55
		c.	5567	2371	8845
		d.	52.14	43.59	32.92
4.	Protharapur	a.	75	53	148
		b.	93.76	44.61	161.80
		c.	5044	1877	6115
		d.	53.79	42.07	37.71
5.	Brischganj	a.	70	51	115
		b.	99.99	42.91	115.37
		c.	5079	1713	3483
		d.	50.79	39.92	30.19
6.	Garacharma	a.	75	72	308
		b.	91.53	71.46	259.28
		c.	3163	2143	7289
		d.	34.55	29.98	28.11
7.	Calicut	a.	103	130	535
		b.	119.97	73.49	340.23
		c.	10366	5818	22452
		d.	86.40	79.16	65.99
8.	Dollyganj	a.	52	41	64
		b.	73.69	40.25	57.16

(Contd.)

TABLE 4.11 *(Contd.)*

1	2		3	4	5
		c.	1482	689	833
		d.	20.10	17.11	14.57
9.	Makkapahar	a.	81	102	536
		b.	99.80	58.90	438.77
		c.	10219	5804	38093
		d.	102.39	98.54	86.81
10.	Shadipur	a.	57	52	240
		b.	84.93	52.90	309.52
		c.	2129	1156	6134
		d.	25.06	22.15	19.81
11.	Nayagaon	a.	60	49	169
		b.	85.49	50.53	160.53
		c.	2181	1043	2802
		d.	25.61	20.64	17.45

a. = Number of aestivating pockets.
b. = Area of aestivating pockets.
c. = Total number of snails in the aestivating pockets.
d. = Average population of snails per m^2.

Being a prolific breeder and each individual being capable of laying eggs because of being hermaphrodite, it soon rehabilitates itself by rebuilding its population, spreads all over again and becomes a menace to vegetation.

5

STRUCTURE AND FUNCTION

Shell

The shell of *Achatina fulica* is ovate — conoid, pale yellowish, streaked transversely with brown, somewhat glossy, more or less irregularly decussated with fine spirals except on the last whorl (Fig. 5.1). The spire is convex, and the suture is impressed, with an acute apex. There are 8 convex whorls, the last equalling nearly half the length of the shell. The aperture of the shell is elliptic ovate, white or pale lilac within, peristome is straight, acute; the margins are connected by a callus on the parietal wall. The columella is slightly concave and truncate (Gude, 1914).

Gude (1914) has mentioned the size of the shell of Calcutta specimens as follows : Length 9.4 cm, maximum diameter 4.5 cm and the aperture 4.8 cm. Peterson (1957b) collected on Guam in 1955, a shell of *Achatina* which was 19.5 cm in length. The author collected a shell of *Achatina* measuring 17.8 cm in length on the Andaman — the largest ever recorded from the Indian region. Lange (1950) found the verage length of the shells on Saipan to be 6.35-9.52 cm. He states that the shell whorls have transverse striations and microscopic spiral striae.

On Saipan, Lange observed the presence of 'thin' "paper-like shells" easily broken by handling, although shells of normal thickness were also recorded from the same area. This variation could be related to genetic factors regulating carbonate metabolism or it may be related simply to the quantity of carbonate in the respective habitats (Lange, 1950).

Kondo (1964) estimated that the shell becomes mature at the age of 11 months, 10 days (324 days) when the shell attains the length of 11.0-12.9 cm, the maximum diameter 5.6-6.3 cm and has 8.5 whorls.

Olson (1973) has reported that on Hawaii, the largest shell found by him in nature had 7.5 whorls and measured 10.61×5.5 cm. When he examined one year old cultured snails, he found that size of the shells varied from 10.5×5.25 cm to 14.4×6.3 cm and the shells had 8.5 whorls which supports the observations of Kondo (1964) on Saipan.

FIG. 5.1: *Achatina*-Shell, about 17.8 cm — the largest size ever reported from the Indian region

Head and Foot

Like other gastropods, *A. fulica* is asymmetrical, the asymmetry being confined to the organs of the visceral mass (Pers. Comm : S.K. Raut 5.11.89).

The head (Fig. 5.2) is situated at the anterior end of the body and has the mouth at its anterior extremity. The head bears two pairs of contractile and invaginable tentacles (Ghose, 1962a). Both the pairs of tentacles are olfactory and tactile but only the larger posterior pair (also called ocular) bears eyes at the apex (Fig. 5.2). The ocular tentacles of Stylommatophora are considered to correspond to the single pair of Basommatophora (Hyman, 1967).

The ocular tentacles with the eyes and olfactory organs and the anterior pair with olfctory organs can regenerate fully (Ghose, 1962a). Time taken in regeneration of the tentacles with the eyes and olfactory organs is less in younger snails and more in older ones.

The foot (Fig. 5.2) is a muscular organ and projects from the ventral surface of the body. It has a flat sole and is used for locomotion. It expands and contracts during locomotion. The locomotion in *Achatina* takes place by crawling/creeping by the alternate expansion and contraction of the foot.

The pedal gland is lodged within the foot along the median line in the large sinus of the foot except for a small portion of the posterior end. It is a cream-coloured, elongated dorsoventrally flattened structure. The dorsal surface of the gland is convex with a longitudinal groove which runs along the mid-ventral surface.

Posteriorly it is narrow somewhat round, extends beyond the middle of the foot and is embedded in the muscles. Anteriorly, the gland is thin and flat and opens at the anterior end of the foot. In the full-grown snails, the size of the gland is 55×2 mm (Ghose, 1963b).

A narrow longitudinally disposed slime canal, running through the gland mass, receives mucus secretion from the walls of the gland and discharges it to the outside through an opening as described above. The mucus discharged on the surface of the foot facilities the gliding movement of the snail.

In addition to the pedal gland, there are a large number of unicellular mucous glands in the foot and these also secrete mucus (Ghose, 1963b).

FIG. 5.2: *A. fulica* — head and foot protruding out of the shell. Eyes distinctly visible on the longer posterior (ocular) tentacles

The Muscular System

Unlike vertebrates, muscles in molluscs are not distinguishable as voluntary and involuntary. However, there is clear physiological distinction and they can be identified as somatic and visceral muscles comparable to some extent respectively to the voluntary and involuntary muscles of the vertebrates (Ghose, 1966).

The entire *columellar muscle complex* comprises somatic muscles. In addition, somatic muscles are also present in the foot, visceral stalk, tentacles, labial palps, buccal mass, penis, vas deferens and oviduct. The visceral muscles are present only in the mantle, alimentary tract and other visceral organs. In *Achatina* both types of muscle fibres are non-striped.

The columellar muscle complex lies to the left of the hind gut hence it is designated as the left columellar muscle. It is divided into two large muscles : the pedal retractor and the free retractor muscle. These lie side-by-side and are attached to the columella close to the junction of the body-whorl and the penultimate whorl.

From the two laterodorsal and posterodorsal surfaces of the foot, ventral to the neck, projects the *pedal retractor muscle* (Fig. 5.3a,b) backward and after gradually tapering for about 30 mm, it becomes spirally twisted and completes one full turn. This portion is attached to the columella and its spiral shape corresponds to the shape of the columella. The function of this muscle is to anchor the body of the snail to the shell. It also seems to help in the contraction of the foot, but its role in this regard appears to be very small.

The free retractor muscles are proximally attached to the columella beyond the origin of the pedal retractor. The proximal 30 mm of the muscle, is spirally twisted and completes 1.25 turns. In this region the muscle is ribbon-like. Distally it divides into right and left retractors.

The right free retractor muscle has five branches running anteriorly to the buccal mass, snout, vas deferens and the anterior part of the foot. These are : the penial, the pedal, the ocular, the dorsolateral and the dorsomedio-buccal retractors.

The penial retractor (Fig. 5.3a,b) arises from the medial side of the free retractor at a distance of about 25 mm from the origin on the columella and runs anterolaterally to end on the hair-pin loop of the vas deferens at the posterior end of the penis (Ghose, 1963d). Quick withdrawal of the penis is affected by the contraction of the penial retractor. In addition to the normal type of insertion described above, the penial retractor may be a separate band unconnected with the free retractor and is inserted either on the right of the diaphragm over the median haemocoelic chamber or it bifurcates into two equal bands at a short distance from the penis or it shifts further to be attached to the left side of the dorsum of the foot near the neck.

There are three *right pedal retractors* (Fig. 5.3) which arise from the right side, anterior to the root of the penial retractor and run anterolaterally. The first one divides into five branches while each of the second and third divide into two or three branches. The branches end on the ventrolateral wall of the foot at short distances along an anteroposteriorly directed line.

The right ocular retractor (Fig. 5.3) originates slightly anterior and ventral from the third pedal retractor. In its forward course it sends a median branch, the cerebral retractor, which unites with its fellow of the other side to form a transverse muscle on the posterodorsal part of the buccal mass. This is connected with the cerebral ganglia by a membranous sheet of tissue (Ghose, 1962 b). The ocular retractor runs forward, become membranous and fan-shaped and divides into two.

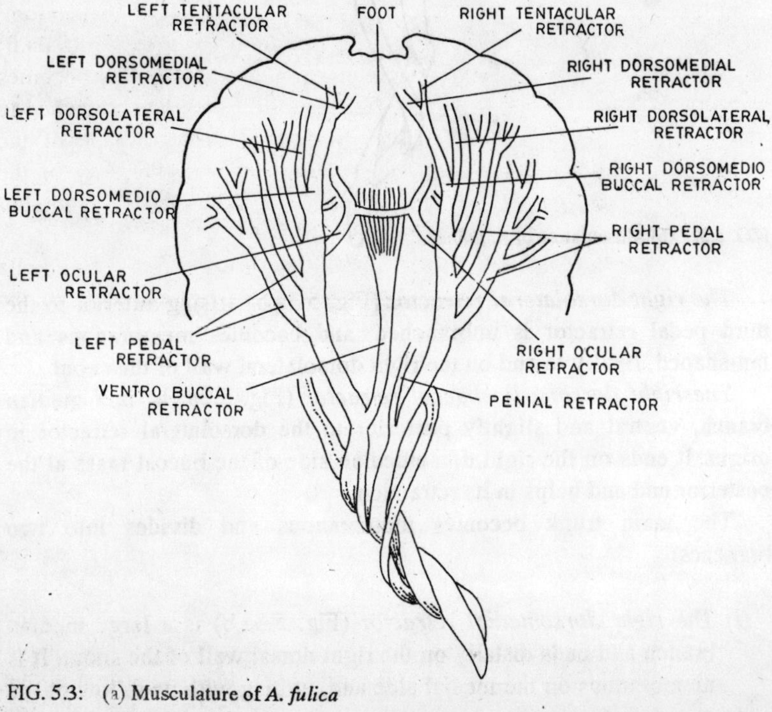

FIG. 5.3: (a) Musculature of *A. fulica*

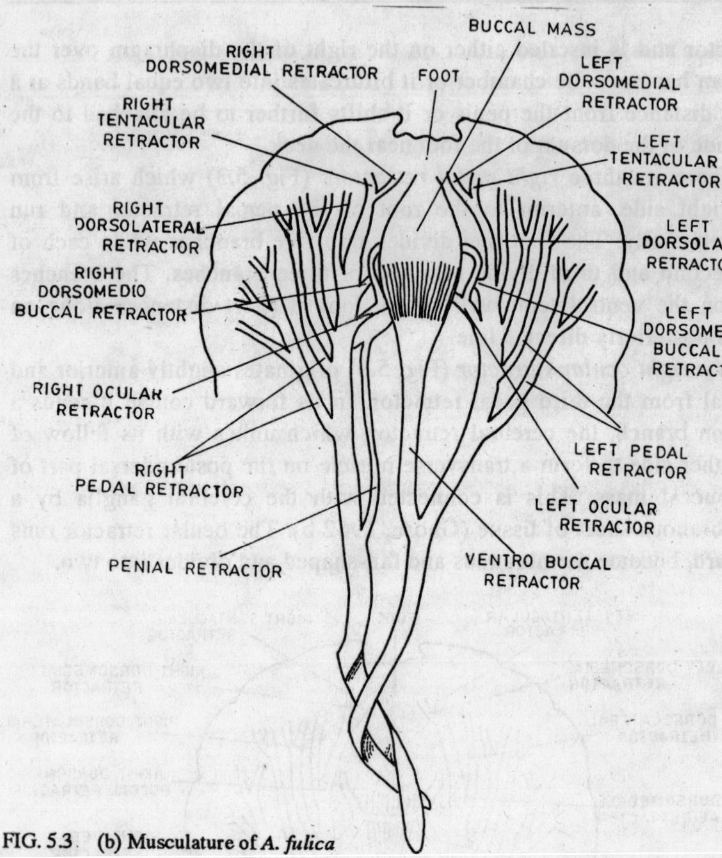

FIG. 5.3 (b) Musculature of *A. fulica*

The right dorsolateral retractor (Fig. 5.3a,b) arising anterior to the third pedal retractor is unbranched, and becomes membranous and fan-shaped, finally to end on the right dorsolateral wall of the snout.

The right dorsomedio buccal retractor (Fig. 5.3a, b) is a median branch, ventral and slightly posterior to the dorsolateral retractor in origin. It ends on the right dorsomedian side of the buccal mass at the posterior end and helps in its retraction.

The main trunk becomes membranous and divides into two branches:

(i) *The right dorsomedial retractor* (Fig. 5.3a,b) is a large median branch and ends distally on the right dorsal wall of the snout. It is membranous on the medial side and, uniting with its fellow of the other side, forms a complete dorsal covering over the buccal mass. Anteriorly, it ends on the tip of the snout and is connected posteriorly with the cerebral ganglia and adjacent nerves. It helps in the retraction of the snout along with the nerve ganglia.

(ii) *The median branch of the right tentacular retractor* (Fig. 5.3) ends on the dorsolateral wall of the snout at its tip and the lateral branch on the right ventral tentacle.

Left Free Retractor Muscle

It is stouter than the right one and has the following branches :

The ventrobuccal retractor (Fig. 5.3) is a stout median branch and separates from the main trunk about 30 mm forward from the origin on the columella. This muscle runs anteromedially, becomes membranous and fan-shaped anteriorly and divides into several branches to end on the ventral surface of the buccal mass, along the posterior margin of the buccal cartilage and anterior to the radular sac.

The left pedal retractors (Fig. 5.3) are two in number and arise from the left side anterior to the ventrobuccal retractor. Their branches and the course correspond with those of the right pedal retractors but the branches of the first one end on the foot slightly posteriorly.

The left ocular retractor (Fig. 5.3) arises from the main trunk anterior to the origin of the second left pedal retractor and runs forward. Its branches and the course correspond to those of the right ocular retractor.

The left anteropedal retractor arises from the left side of the main trunk close to the origin of the left ocular retractor and soon divides into two branches : a median branch which runs forward and, being membranous ends on the anterodorsal wall of the snout, towards the left and the second branch which runs laterally, becomes membranous and ends on the ventrolateral wall of the foot.

The main trunk of the left free retractor becomes membranous. Its subsequent branches and their course are very similar to those of the corresponding branches of the right one.

Due to the contraction of the powerful columellar muscles, the anterior tentacles, ocular tentacles, snout and the anterior part of the foot are retractable. The retraction begins at the tip of the snout and gradually runs backwards. With the contraction of the columellar muscles, the intrinsic muscles of the foot also contract and the foot becomes shortened and wavy in cutline. The visceral stalk contracts due to the contraction of its longitudinal muscles. By a coordination of the columellar muscles and the muscles of the visceral stalk and foot, the snail, is able to withdraw itself rapidly into the shell.

The Digestive System

The digestive system of *A. fulica* has been worked out in detail by Ghose (1963c). It is "remarkably simple" and consists of (a) an alimentary canal, (b) the salivary glands and (c) the digestive gland.

The Alimentary Canal

It is divisible into (i) the mouth, (ii) the buccal mass, (iii) the oesophagus, (iv) the crop, (v) the stomach, (vi) the intestine, and (vii) the rectum (Fig. 5.4).

(i) *The Mouth:* It is a semi-circular opening located at the anterior end of the body with a plicate margin and enclosing a small vestibule posteriorly. A pair of labial palps project laterally arising slightly dorsally from each corner of the mouth.

(ii) *The Buccal Mass:* It is thick and highly muscular structure, narrower anteriorly and broader posteriorly (Fig. 5.4). The anterior border is bound dorsally by the jaw and ventrally by the sphincter muscle. It can be protruded forward by the contraction of the protractor muscles and retracted partly by the relaxation of the protractor muscles and partly by the contraction of the retractor muscles (Ghose, 1963c).

The jaw is a semilunar, homogeneous cartilage, about 6.25 mm long. The anterior surface is capped with a thick layer of brownish plicate cuticle whereas the posterior surface is deeply concave. The sphincter muscle is attached at the concavity. The vertical height of the jaw is greater than its thickness. The cartilage continues posteriorly from the ventral border and gradually becomes thicker.

The buccal cartilage is a thick, horse shoe-shaped structure with a broad body and the limbs gradually coming closer and nearly touching each other at their ends. The deeply concave face of the body is directed posterodorsally while the convex face is anteroventral. The limbs are long, broad, concavo-convex and slightly thickened towards their free ends. The anterior end of the radula is placed on the anterior

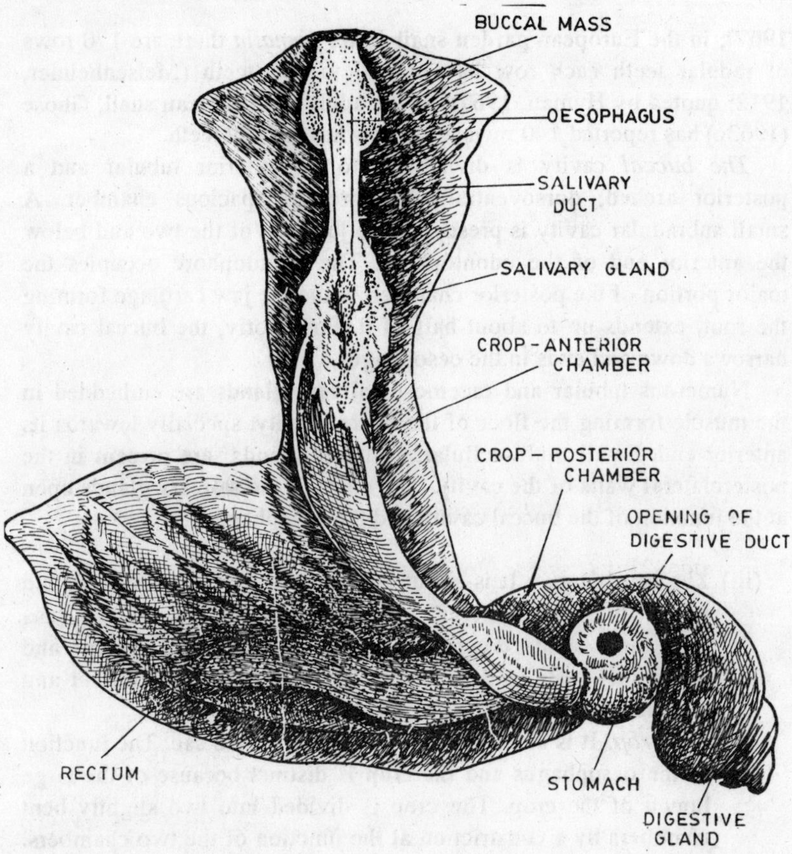

FIG. 5.4: Digestive organs of *A. fulica*

convex surface of the cartilage. The elastic membrane, the radular and subradular retractors meet the buccal cartilage at the free ends of its arms, while the sphincter muscle is inserted into the side of the arms.

The radula is elongated and has a narrow anterior and a blunt posterior end. The anterior end is recurved and lies against the anterior convex face of the buccal cartilage. Its sides at the posterior end become folded to enclose the subradular collostyle.

In Stylommatophora the rows of radular teeth and the number of teeth per row is often large. Thus in the slug *Arion empiricorum* there are 160 rows with 100 teeth per row (Guiart, 1900; quoted by Hyman,

1967); in the European garden snail *Helix pomatia* there are 170 rows of radular teeth each row having 140 to 150 teeth (Meisenheimer, 1912; quoted by Hyman, 1967). In the adult giant African snail, Ghose (1963c) has reported 140 rows each row having 129 teeth.

The buccal cavity is divisible into an anterior tubular and a posterior arched, dorsoventrally compressed spacious chamber. A small subradular cavity is present at the junction of the two and below the anterior end of the odontophore. The odontophore occupies the major portion of the posterior chamber, while the jaw cartilage forming the root, extends up to about half of it. Posteriorly, the buccal cavity narrows down and ends in the oesophagus.

Numerous tubular and racemose salivary glands are embedded in the muscle forming the floor of the buccal cavity, specially towards its anterior end. Small multi-cellular, scattered glands are present in the posterolateral walls of the cavity. The ducts of the salivary glands open at the junction of the buccal cavity and the oesophagus.

(iii) *The oesophagus:* It is about 36 mm long, narrow, thick-walled tube with several internal longitudinal folds (Fig. 5.4). It runs backwards and downwards from the buccal mass and posteriorly ends in the crop at the junction of the anterior and median haemocoelomic chambers.

(iv) *The crop:* It is about 50 mm long, thin-walled sac. The junction of the oesophagus and the crop is distinct because of the large lumen of the crop. The crop is divided into two slightly bent chambers by a constriction at the junction of the two chambers. The anterior chamber (Fig. 5.4) is longer than the posterior one being about 37 mm long and is broadest near its anterior end while posteriorly it narrows, becomes thick-walled and the constriction acts as a sphincter to check free flow of substances from the anterior to the posterior chamber. The posterior chamber (Fig. 5.4) opens in the stomach on the right side by a round aperture. A short and broad duct from the anterior lobe of the digestive gland opens in the posterior chamber of the crop, a little before its opening in the stomach.

(v) *The stomach :* It lies embedded in the mass of digestive gland (Fig. 5.4). It is more or less heart-shaped with its round posterior end shifted slightly towards the right. Ventrally, the anterior border is almost straight but dorsally it is depressed in the middle and bears two round forward prolongations. A large duct from the posterior lobe of the digestive gland opens on the ventral surface of the stomach.

(vi) *The intestine:* It is a 75 mm long, thin-walled tubular structure arising from the left anterior border of the stomach. Anteriorly, the intestine runs straight for a very short distance, then turning to left, it forms an arch and proceeds along the posterior end of the kidney. It then takes a turn slightly to the right and dorsally gets completely embedded in the mass of the digestive gland. Later running anterolaterally along the left border of the visceral mass, it ends in the rectum.

(vii) *The rectum:* It is a very thin walled tube about 68 mm long (Fig. 5.4) running along the posterior border of the mantle cavity. The wall of the posterior part for about 6 mm has strong longitudinal folds which obliterate its lumen and act as a spincter. The rectum opens to the outside through the anus.

The Salivary Glands

These are cream-white, paired, multilobed structures (Fig. 5.4), each being about 27 mm in length. These lie dorsolaterally on the posterior end of oesophagus and dorsally on the anterior part of the crop.

The glands are narrow anteriorly and relatively broader posteriorly. A long salivary duct measuring about 25 mm arises from the anterior end of each gland and opens in the buccal cavity.

Each lobe of the gland is sub-divided into a number of lobules; each lobule is enclosed in a thin capsule and is made of a large number of alveoli; from each alveolus, emerges a narrow duct, and all the ducts of the alveoli unite to form as a single large duct. However, the ducts emerging from different lobules open "either directly in the main duct" of the gland or "unite with ducts coming from other lobules".

The Digestive Gland

It is a dull, blackish-brown, large, thick, bilobed, spirally-twisted structure. Posteriorly, it narrows and occupies nearly the whole of the apical visceral mass (Fig. 5.4). The anterior lobe is short, broad, thick and extends about half way to the stomach. The posterior lobe is more or less conical, broad and thick anteriorly and narrower and spirally-twisted posteriorly. The gland is made up of a large number of lobules of varying thickness and several of these unite to form interlobular ducts which unite with each other and move forward. The lobules are full of dull-brown secretion produced from the gland cells.

The Circulatory System

Generally the circulatory system of pulmonates is comparatively less open than in other gastropods.

The circulatory system of pulmonates has been studied in a number of species, some of the species being *Lymnaea stagnalis, Ferrissia* sp. (Basommatophora), *Arion empiricorum* and *Helix pomatia* (Stylommatophora) (Hyman, 1967). Ghose (1964a) has studied the structure and functioning of the heart of *A. fulica.*

The pericardium and the heart: The pericardial chamber is a closed sac except for the reno-pericardial aperture and lies transversally in the left side of the mantle cavity in its anterior part parallel to the kidney. It is elongate-oval and there is a narrow space between the auricular part of the chamber and the kidney. The ventricular part is in contact with the wedge-shaped forward process of the kidney. Dorsally, the pericardium is fused with the mantle but ventrally it is free.

The chamber is narrower at the ventricular end (Fig. 5.5) where it is fused with the apex of the ventricle and the base of the aortic ampulla. It is broader and somewhat rounded at the auricular end and is fused with the narrow end of the auricle and the base of the pulmonary vein.

The pericardium (Fig. 5.5) is a tough membrane, traversed by numerous lacunae. It is bound on both sides by one cell thick epithelium of flat cell. A thin sheet of longitudinal muscle underlies the epithelium and encloses a thick, compact muscular layer of interwoven fibres.

The pericardial cavity is in direct communication with the lumen of the kidney through the ciliated reno-pericardial passage. The flow of the fluid from kidney into the pericardial chamber is prevented due to the one-way movement of the cilia of the reno-pericardial passage.

The auricle (Fig. 5.5) is wine-glass-shaped and thin-walled. Its apex is narrower and continuous with the pulmonary vein. The broad concave base receives the convex base of the ventricle, and the two proceed in a narrow circular area round the auriculo-ventricular orifice. The thin wall of the auricle is strengthened with longitudinal, transverse, and oblique muscular strands attached to its inner surface and also running through the cavity. These strands are branched, and often fused at the points of their intersections, forming a network, and occupying a considerable portion of the auricular cavity; the size of the cavity is consequently reduced.

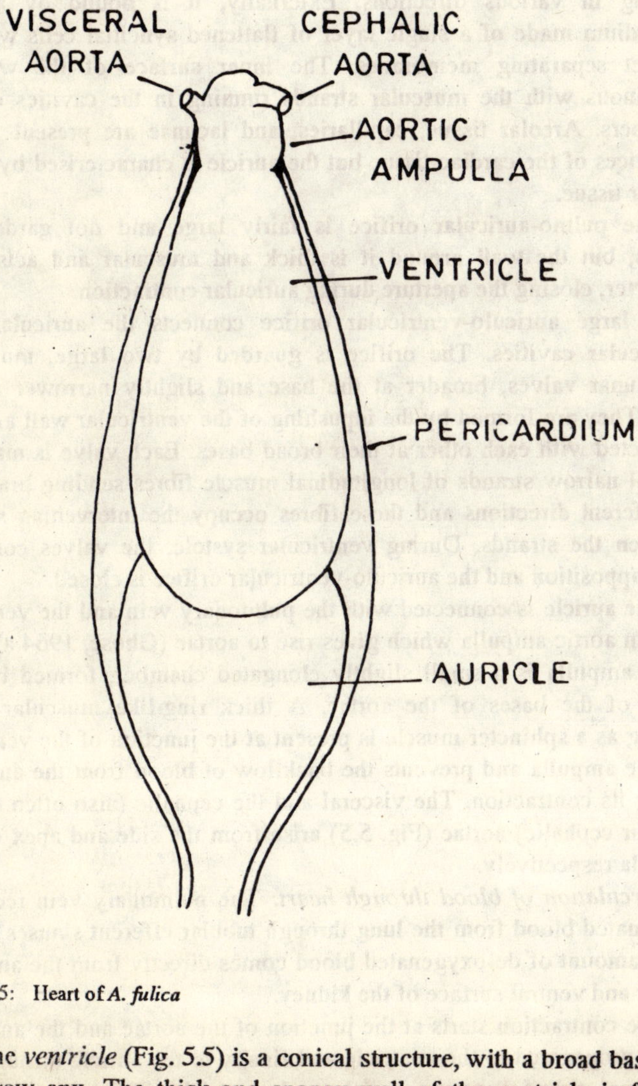

FIG. 5.5: Heart of *A. fulica*

The *ventricle* (Fig. 5.5) is a conical structure, with a broad base and a narrow apx. The thick and spongy wall of the ventricle is further strengthened by variously-oriented muscles, arranged in trabeculae and forming prominent ridges on the inner surface. Branches of muscular strands like those of the auricle are also present and being numerous and thicker, they greatly reduce the size of the cavity of the ventricle.

The *myocardium* constitutes the walls of both auricle and ventricle and is formed by closely-packed non-striated, ramified muscle fibres, running in various directions. Externally, it is bound by a thin epicardium made of a single layer of flattened syncitial cells without distinct separating membranes. The inner surface of the wall is continuous with the muscular strands running in the cavities of the chambers. Areolar tissue, capillaries, and lacunae are present in the interstices of the cardiac tissue but the auricle is characterised by more areolar tissue.

The pulmo-auricular orifice is fairly large and not garded by valves, but the wall around it is thick and muscular and acts as a sphincter, closing the aperture during auricular contraction.

A large auriculo-ventricular orifice connects the auricular and ventricular cavities. The orifice is guarded by two large, muscular semi-lunar valves, broader at the base and slightly narrower at the apex. They are formed by the inpushing of the ventricular wall and are connected with each other at their broad bases. Each valve is made of several narrow strands of longitudinal muscle fibres sending branches in different directions and these fibres occupy the intervening spaces between the strands. During ventricular systole, the valves come in close apposition and the auriculo-ventricular orifice is closed.

The auricle is connected with the pulmonary vein and the ventricle with an aortic ampulla which gives rise to aortae (Ghose, 1964 a). The aortic ampulla is a small slightly elongated chamber, formed by the union of the bases of the aortae. A thick ring-like muscular band serving as a sphincter muscle is present at the junction of the ventricle and the ampulla and prevents the backflow of blood from the ampulla during its contraction. The visceral and the cephalic (also often called anterior cephalic) aortae (Fig. 5.5) arise from the side and apex of the ampulla respectively.

Circulation of blood through heart: The pulmonary vein receives oxygenated blood from the lung through tubular efferent sinuses and a small amount of de-oxygenated blood comes directly from the anterior border and ventral surface of the kidney.

The contraction starts at the junction of the aortae and the ampulla and is propagated to the ventricle and finally to the auricle through the junctional tissue between auricle and ventricle. There is no pause between the contraction of auricle and ventricle and the wave of contraction passes so rapidly that different components of the heart appear to contract simultaneously. Transmission of contraction by peristalsis is entirely absent. The back-flow of blood from the auricle to the pulmonary vein and the aortic ampulla to the ventricle is prevented by the closure of the pulmo-auricular and ventirculo-ampullar orifice respectively brought about by the contraction of their muscular walls. The auriculo-ventricular valves stop back flow from the ventricle to the auricle. The contraction of aortic ampulla helps the ventricle in driving blood to the aortae with greater force.

The heart beats rhythmically. In an average-sized specimen the rate of heart beat is approximately 25 to 30 per minute. It may come down to 9 to 10 per minute during aestivation and increases considerably when the animal is under mechanical, chemical and electrical stress.

No published record of detailed arterial and veinous systems of *A. fulica* is available.

In pulmonate gastropods to which *A. fulica* also belongs, generally the arterial blood vessels are well-defined arteries which finally terminate into capillaries. The existence of a network of capillaries in pulmonates was discovered as early as in 1883 by Nalepa (quoted by Hyman, 1967). Network of capillaries occur on the surface or in the interior of all the organs. Probably the capillaries open into transitional channels through contractile openings; these transitional channels discharge blood into the large venous sinuses. In pulmonates, like other gastropods, the interior spaces are just blood sinuses and not coelomes; true coelomes are only the cavities of the pericardial sac and around the ovotesties (Hyman, 1967).

The pulmonate veinous blood vessels are the centrally-placed cephalic and visceral sinuses, right and left marginal sinuses, rectal sinus along the rectum and one median and paired lateral pedal sinuses in the foot. Finally the entire veinous blood reaches a circular channel, the veinous circle along the edge of the pulmonary sac including a ring around the pneumostome. A large number of efferent pulmonary arteries from the veinous circle branch into capillaries which form a network in the roof of the pulmonary sac in which the blood is aerated. Numerous efferent pulmonary veins join a main pulmonary vein which opens into the adjacent auricle. It seems that in the pulmonary net, there is no direct passage from the arterial capillaries to the veinous capillaries. Efferent capillaries adjacent to the nephridium, end in the nephridial sinuses and are then gathered together to enter the pulmonary vein before it discharges into the auricle. Thus, it is evident that the heart carries only aterial blood some of which has also been made free from excretary matter (Hyman, 1967).

The Respiratory System

Like other pulmonates, *Achatina* also is an air breather. In pulmonates having an outer protective shell like *Achatina,* the edge of the mantle is fused to the snail's neck except for a small opening on the right side. This opening is known as pneumostome which is contractile (Hyman, 1967). Like other pulmonates, in *Achatina* also, gills are absent. The mantle cavity is transformed into a pulmonary sac which acts as a respiratory chamber. This chamber is highly ramified with blood vessels on the inner surface of the roof. It is filled with air by the snail's rising to the surface and opening of the pneumostome. It is by this opening that the snail communicates with the external medium.

In *A. fulica,* during aestivation/hibernation also, when the aperture of its shell is closed by a thin, fragile epiphragm, there is provision for respiration. The epiphragm has a small slit over the pneumostome to allow reduced respiration (Mead, 1961).

Detailed published account of the respiratory system of *A. fulica* is not available.

The Sensory Organs

The sensory organs in *A. fulica* have been described by Ghose (1962a). These comprise (a) organs of sight, (b) olfactory organs, (c) organs of equilibrium and (d) tactile organs.

(a) *Organs of Sight*

Eyes of *A. fulica* are lodged slightly dorsally at the tip of the two ocular (posterior) tentacles and look-like round black dots.

The eye-stalk is bound externally by a homogenous lamina covering a columnar epithelium. The lamina and the epithelium are continuous over the eye to form the outer cornea which is supported below by a very thin layer of fibrous connective tissue. A narrow posterior strip of the fibrous layer runs posterolaterally and medially to form a capsule or the eye vesicle. The inner cornea lies below the fibrous layer and is restricted to the anterior part of the eye in the region of the outer cornea only. At the base of the eye is the retinal layer which is lined internally by a very thin membrane and a large number of closely-placed cone-shaped structures. The optic nerve pierces the eye vesicle posteriorly and forms a nerve layer inside the capsule, the nerves finally ending in the retinal cells. The round space enclosed by the layer of cones is filled up with a homogeneous, structureless lens.

(b) *Olfactory Organs*

There are four olfactory organs each lodged in an anterior and posterior tentacles.

Each olfactory organ consists of a club-shaped olfactory ganglion in the centre of which are scattered nerve cells, whereas the periphery of the ganglion is crowded with nerve cells. From the anterior and anterolateral sides of the ganglion, slender processes arise and ramify on the covering epithelium of the tentacles. Ganglion of the ocular tentacle is relatively larger (Ghose, 1962a).

The ocular tentacles (Fig. 5.2) with the eyes and olfactory organs and the anterior tentacles with the olfactory organs if cut, are capable of complete regeneration. The time taken in regeneration depends upon factors like age of the snail and the season. Tentacles in younger snails, during favourable season regenerate comparatively more quickly. Also if the antenna is cut from near the apex it regenerates more quickly than when cut from near the base. However, Ghose (1962a) has observed that even without the eyes and olfactory organs, the snail "behaves quite normally and possibly the animals do not feel much inconvenience due to want of eyes and olfactory organs".

(c) *Organs of Equilibrium*

In the snail, the organs of equilibrium are the statocysts which enable it to know its position in space. These are two in number. Each is situated on the pedal ganglion dorsolaterally, slightly posterior to its middle. The statocyst is a round sac bound externally by a thin layer of fibrous tissue enclosing a cavity. The cavity is full of a fluid containing about 50 elliptical statoliths (statoconia). The statocyst is supplied by a nerve from the cerebral ganglion of the side.

(d) *Tactile Organs*

All the four tentacles of the snail also act as tactile organs. In addition, the anterior ends of the foot and snout also are sensitive to touch.

The Nervous System

The nervous system of *A. fulica* has been worked out by Ghose (1962b). It consists of several pairs of nerve ganglia and nerves emerging from these. The main ganglia are cerebral, buccal, pleural, parietal, abdominal and pedal. The abdominal ganglion is single, all others are paired. The buccal and pleural pairs remain separate while the rest of the pairs get fused during development, each pair forming a single ganglionic mass.

The Cerebral Ganglia

The cerebral ganglia (Fig. 5.6) are paired, almost spherical structures, broader towards their outer and narrower towards the inner ends. The cerebral ganglia are "chiefly sensory". They are fused towards their inner ends to form a ganglionic mass which lies dorsal to the oesophagus, about 18 mm behind the buccal mass. The cerebral ganglia supply nerves to the head and anterior region of the visceral stalk and the organs lodged there. From each ganglion arise a number of nerves which can be arranged according to their origin into : (a) Dorsal antero-median nerves and (b) Ventral antero-lateral nerves.

Dorsal antero-median nerves

The following nerves arise from the dorsal and antero-median aspects of each cerebral ganglion:

1. *Ocular nerve:* It is slender and innermost in origin, it runs anterolaterally to join the ocular retractor muscle.
2. *Ommatophoral nerve:* It is stoutest of the group and arises anterolaterally to the ocular nerve, runs anterolaterally and making a broad arch, becomes inserted into the ocular retractor muscle in front of the ocular nerve. Slightly posterior to the tip of the ocular tentacle the nerve bifurcates; the slender dorsal branch ends in the eye, while the stout ventral branch swells to form a club-shaped olfactory ganglion sending processes to the tip of the

ocular tentacle.

3. *Superior frontal nerve:* It is anterior and slightly ventral to the ommatophoral in origin, runs anterolaterally and dorsally to end in the snout at the base of the ocular tentacle.
4. *Inferior frontal nerve:* It is a slender nerve, anteroventral to the superior frontal and runs almost parallel to it to end in the dorsal and dorsolateral wall of the snout.

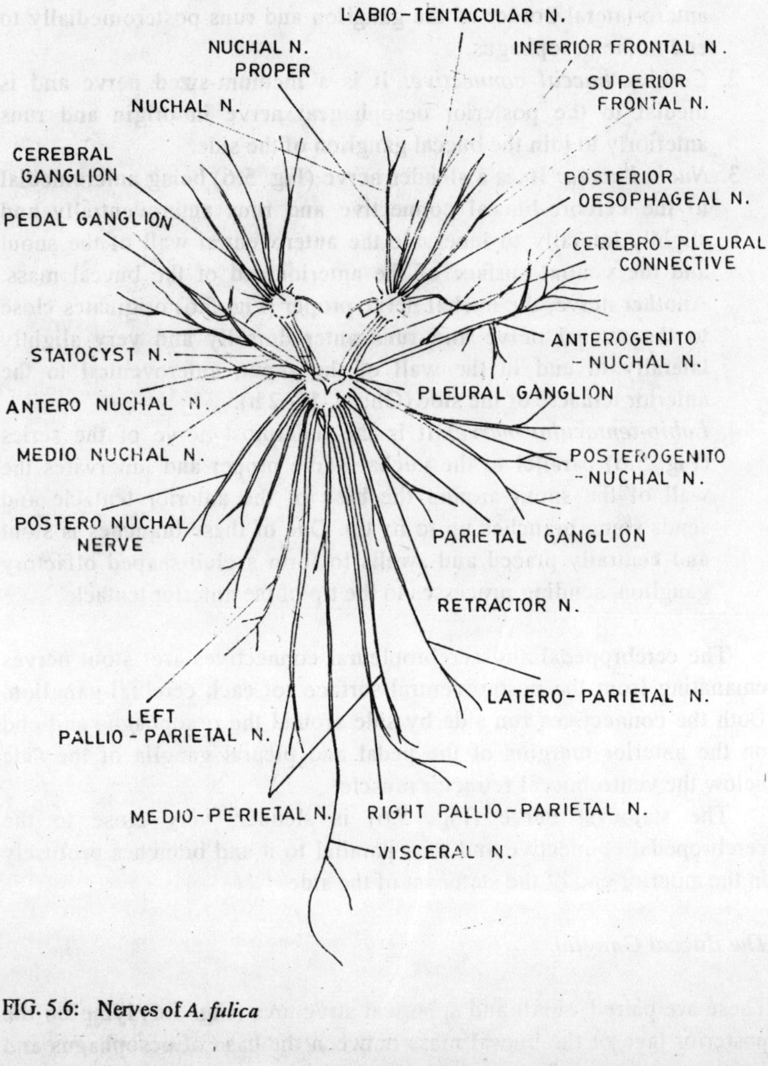

FIG. 5.6: Nerves of *A. fulica*

Ventral Antero-lateral Nerves

The following nerves arise from the ventral and antero-lateral side of each cerebral ganglion:

1. *Posterior oesophageal nerve :* It is a small and slender nerve which arises from the ventral surface of the outer and antero-lateral border of the ganglion and runs posteromedially to end in the oesophagus.
2. *Cerebro-buccal connective:* It is a medium-sized nerve and is medial to the posterior oesophageal nerve in origin and runs anteriorly to join the buccal ganglion of the side.
3. *Nuchal nerve:* It is a slender nerve (Fig. 5.6) being anteromedial to the cerebro-buccal connective and runs anteroventrally and slightly laterally to innervate the anteroventral wall of the snout and the ventral surface of the anterior end of the buccal mass. Another nerve, the nuchal nerve proper (Fig. 5.6) originates close to the nuchal nerve and runs anterodorsally and very slightly laterally to end in the wall of the snout, anteroventral to the anterior tentacle of the side (Ghose, 1962 b).
4. *Labio-tentacular nerve:* It is the innermost nerve of the series (Fig. 5.6) parallel to the nuchal nerve proper and innervates the wall of the snout around the base of the anterior tentacle and sends some branches up to its tip. One of these branches is stout and centrally placed and swells to form a club-shaped olfactory ganglion, sending processes to the tip of the anterior tentacle.

The cerebropedal and cerebropleural connectives are stout nerves emanating from the posteroventral surface of each cerebral ganglion. Both the connectives run side by side around the oesophagus and end on the anterior margins of the pedal and pleural ganglia of the side below the ventrobuccal retractor muscle.

The statocyst nerve (Fig. 5.6) is slender, very close to the cerebropedal connective and runs parallel to it and branches profusely in the anterior end of the statocyst of the side.

The Buccal Ganglia

These are paired, small and spherical structures (Fig. 5.6) lying on the posterior face of the buccal mass between the base of oesophagus and the radular sac. These are connected by a narrow transverse commisure. The cerebrobuccal connective joins on their outer side. These ganglia are both sensory as well as motor and supply nerves to the buccal mass.

Both anteriorly and posteriorly several nerves arise from each buccal ganglion and innervate the organs of the side.

Four nerves arise from its anterior side. These are:

1. *Lateral buccal nerve:* Originating close and anterior to the cerebro-buccal connective, it runs anterolaterally (Fig. 5.6) to reach the border of the buccal mass. It proceeds further anteriorly along the lateral side of the buccal mass to enter the sphincter muscle. Its subsequent course up to the anterior end is within the buccal mass.
2. *Dorsolateral buccal nerve:* It originates from the middle of the root of the lateral buccal nerve (Fig. 5.6) and runs anteriorly and slightly lateral and forward to enter the buccal mass by the side of the oesophagus.
3. *Anterior buccal nerve:* It originates from the middle of the dorsolateral buccal nerve (Fig. 5.6), runs anteriorly and enters the buccal mass by the side of the salivary duct.
4. *Anterior oesophageal nerve:* It is the shortest and the innermost nerve (Fig. 5.6) of the anterior series and runs aneromedially and dorsally, to end on the dorsal surface of the oesophagus very near to its origin.

Nerves arising from the posterior side are:

1. *Posterior buccal nerves:* These fine nerves arise from the posterolateral margin of the ganglion close to each other (Fig. 5.6). They run ventrally and slightly lateral, to reach the ventral surface of the buccal mass and then run anteriorly for a short distance.
2. *Odontophoral nerve:* This fine nerve (Fig. 5.6) lies medially to the posterior buccal nerve, runs ventrally and innervates the odontophore.
3. *Oesophageal and salivary duct nerve:* It arises from the posteromedial border of the ganglion (Fig. 5.6) and bifurcates into two branches. The oesophageal branch which is medial, stout and long and runs posteriorly along the crop up to its end. The

right nerve runs along the dorsal, and the left along the ventral surface of the crop. The salivary duct branch is slender and lateral, and joints the duct of the salivary gland.

The Ganglionic Ring

The paired parietal, pleural and pedal ganglia and the single abdominal ganglion are united together to form a ring, through which the cephalic aorta passes. The two parietal ganglia and the abdominal ganglion form a heart-shaped visceral ganglionic mass, with a slight anterior notch in the middle. This constitutes the dorsal part of the ring. The posteromedian portion of this mass is occupied by the abdominal ganglion with which are also fused the two parietals. The pleurals are spherical and joined with the parietals at their anterolateral ends through a rather distinct connective. The pedal ganglionic mass forming the ventral part of the nerve ring, is a broad transverse mass with a depression at the middle of the anterior border, while an extremely fine groove demarcates the two ganglia on the posterior border. The pedal ganglia are chiefly motor, whereas the pleural and parietal ganglia and the unpaired abdominal ganglion are both motor and sensory.

The pleural ganglia (Fig. 5.6) supply nerves to the middle region of the visceral stalk and some of the genital organs.

Five nerves emanate from the pleural ganglion, two from its right and three from its left side.

The two nerves emanating from the right pleural ganglion are:

1. *Anterogenito-nuchal nerve:* Arising from the anterolateral border of the ganglion (Fig. 5.6) and proceeding anterolaterally, this nerve crosses the second pedal retractor muscle dorsally. It gives off a short branch, which runs anteriorly and innervates the visceral stalk, posterior to the ocular tentacle and the genital atrium; reaching close to the anterior end of the penis, its main trunk bifurcates. The anterior branch ends in the visceral stalk and the penis further forward; the other branch runs posteriorly along the concave border of the penis and gives off a fine branch to the basal vagina and terminates in the penis and the vas deferens.
2. *Posterogenito-nuchal nerve:* Arising from the posterolateral aspect of the ganglion, this nerve follows a posterolateral course,

and then, like the anterogenito-nuchal nerve bifurcates (Fig. 5.6). The anterior branch passes over the genital ducts and ends in the visceral stalk posterior to the penis. The posterior branch gives off a fine nerve to the vas deferens, apical vagina and visceral stalk. The main trunk ends further back.

The three nerves emanating from the left pleural ganglion are:

1. *Anteronuchal nerve:* Originating from the anterolateral aspects of the ganglion (Fig. 5.6), it runs anterolaterally and after crossing the anteropedal retractor muscle dorsally, it ends in the visceral stalk, posterior to the ocular tentacle.
2. *Medionuchal nerve:* This nerve (Fig. 5.6) arises from the lateral border of the ganglion, runs laterally and slightly anteriorly and crossing the second pedal retractor muscle dorsally, it ends in the visceral stalk posterior to the anteronuchal nerve.
3. *Posteronuchal nerve:* The posteronuchal nerve (Fig. 5.6) arises from the posterolateral corner of the ganglion. It runs posterolaterally and bifurcates. The short anterior branch ends in the wall of the posterior part of the visceral stalk while the posterior branch runs further back and ends in the dorsolateral part of the visceral stalk near its junction with the mantle collar.

Nerves of the Visceral Ganglionic Mass

All the nerves arising from the visceral ganglionic mass (Fig. 5.6) run posteriorly and supply the visceral stalk , mantle, free retractor muscles and the visceral organs. Three nerves arise from the right side, five nerves from the left and a stout visceral nerve from the posterior side.

Nerves arising from the Right Side

1. *Right retractor nerve:* It is smallnerve arising from the anterolateral border and runs posteriorly to join the right free retractor muscle.
2. *Lateroparietal nerve:* It is a stout nerve and arises from the posterolateral border of the ganglion, runs posteriorly and slightly laterally, giving off a fine nerve to the cephalic aorta on the left, and bifurcates while passing over the genital ducts. Both the branches end very close to each other, at the junction of the mantle collar and visceral stalk of the side.

3. *Medioparietal nerve:* This nerve arises from the ventral surface of the ganglion close to the origin of the lateroparietal nerve and runs posteriorly to end at the junction of the visceral stalk and mantle.

Nerves arising from the Left Side

1. *Left retractor nerve:* It corresponds to the right retractor nerve in origin and course.
2. *Latero-parietal nerve:* It arises slightly behind the retractor nerve, and runs posterolaterally to end in the left wall of the visceral stalk near its base.
3. *Left pallio-parietal nerve:* Arising from the posterolateral border of the ganglionic mass slightly dorsally it runs posteriorly and slightly laterally up to the junction of the visceral stalk and the mantle. Giving off a few branches to this region, it proceeds towards the left along the junction of the visceral stalk and the mantle to innervate the left half of the mantle.
4. *Medioparietal nerve:* Originating from the posterolateral border of the mass, it runs posteriorly, enters the groove in the middle of the pedal retractor muscle, ventral to the free retractors, and innervates the visceral stalk and the pedal retractor muscle.
5. *Right pallio-parietal nerve:* It originates from the ventral surface of the mass near its posterior border, slightly left to the middle of the mass. Shortly after its origin, it runs to the right, beneath the visceral nerve and runs by the side of the right, medio-parietal nerve to enter the junction of the visceral stalk and the mantle. Giving off a few branches around the point of entrance, it proceeds towards the right along the junction of the visceral stalk and the mantle to innervate the right half of the mantle.

The Visceral Nerve

This large and stout nerve originates from its postero-median and slightly dorsal aspect. It runs posteriorly, and giving off a nerve to the cephalic aorta on the left, it pierces the transverse septum dorsal to the spermathecal sac, enters the posterior haemocoelomic chamber and proceeds along its left wall. On leaving the chamber at the apical end, the nerve runs up to the apical end of the digestive gland along its inner concave margin.

In the posterior haemocoelomic chamber, the nerve gives off two short branches successively on the left and right to supply the left and the right wall of the chamber respectively. It also gives off a short uterine nerve posteriorly to the apical uterus near its basal end. Shortly before its exit from the chamber, it bifurcates into an inner and an outer branch. The inner branch or the intestinal nerve runs along the ventral surface of the intestine for about 25 mm, gives out several branches to it and then reaches the stomach, where it branches profusely. This nerve now runs up to the apical end of the digestive gland, giving off branches to it. The outer branch runs posteriorly and towards the right side and gives off a renal nerve to the left, supplying the kidney, a fine nerve to the heart and an ovotestis nerve which runs along the ovotestis duct to enter the ovotestis. The main trunk now divides into several albumen gland nerves and an apical uterine nerve to supply these organs respectively.

Nerves of the Pedal Ganglia

Nerves from the pedal ganglia (Fig. 5.6) supply exclusively the foot. From the lateral sides of each ganglion, arise nine nerves, some of which run anterolaterally, others only laterally, some others posterolaterally and a few posteriorly. The first nerve comes off the anterior border, slightly medial to the anterolateral corner. It is stout and largest of the first four arising from the pedal ganglia, runs anterolaterally and gives our branches to the anterior and anterolateral end of the foot. The second and the third nerves originate laterally from the anterior half of the ganglion and the fourth more or less from the middle of the lateral side. As we pass from the first to the fourth nerve, these become successively shorter, the fourth one being the shortest of the series. The anterolateral half of the foot is supplied by the nerves of this series.

The fifth nerve or the first nerve of the posterolateral series is the smallest, and runs posterolaterally. The fifth, sixth and seventh nerves are small, dividing repeatedly to form regular plexuses in the sole of the foot. The eighth and ninth nerves are long, the latter being the longest and stoutest of the series. After its origin from the posterior border of the ganglion, the ninth nerve runs posteriorly and very slightly laterally. Beyond the middle of the foot it turns inwards and reaches the posterior end of the foot very near its middle. The eighth nerve runs almost parallel to the ninth, and these two supply the most posterior part of the foot.

The Excretory System

Amongst the invertebrates, Mollusca is the only phylum in which excretory organs are represented by well developed kidneys. Lloyd (1910) has described the kidney of *A. fulica.* Ghose (1964 a) has worked out anatomy of its kidney in detail and has also tried to correlate its structure and functions. The following account of the excretory system of *A. fulica* is based on Ghose's work.

The kidney is lodged in the posterolateral part of the mantle cavity, adhering to the mantle and lying parallel to the heart. Its anterior part is located medially and dorsally while the rest of the organ lies on the left side and is directed ventrally. It is a dull-greyish, single-lobed, soft, flattended, elongated, spongy and transversely-oriented structure (Fig. 5.7). Its average size in the fullgrown snail is about 5×1.2 cm. It consists of numerous elongated folds, enclosing a small lumen in the central part.

Anteriorly it is relatively narrow and round; posteriorly it gradually becomes broader and crescent-shaped end lies against the coil of intestine. The anterolateral border is thin and smooth and is attached to the pulmonary vein, about 1.8 cm away from the anterior end.

In *A. fulica,* the ureter has two limbs, the adrenal and the adrectal. The adrenal ureter (Fig 5.7) runs along the posterolateral border in a deep groove of the kidney, except for about 1.2 cm at the posterior end and about the same distance beyond the posterior limit of the mantle cavity.

A short triangular process wedged in between the ventricular end of the pericardial chamber and the coil of the intestine projects from the anterior corner of the posterior end. In this region, the capsule of the kidney is fused with the pericardium, leaving a small aperture, the reno-pericardial aperture. This aperture leads to a short, narrow passage which after a short distance becomes branched in the mass of the kidney. These branches are ciliated and open in the lumen of the kidney.

The adrectal ureter (Fig. 5.7) is about 75 mm long. On leaving the kidney it takes a sharp turn to the left, making a very narrow arch and reaches the rectum at its beginning and proceeds along its dorsolateral wall to open in the pneumostome, close to the anus.

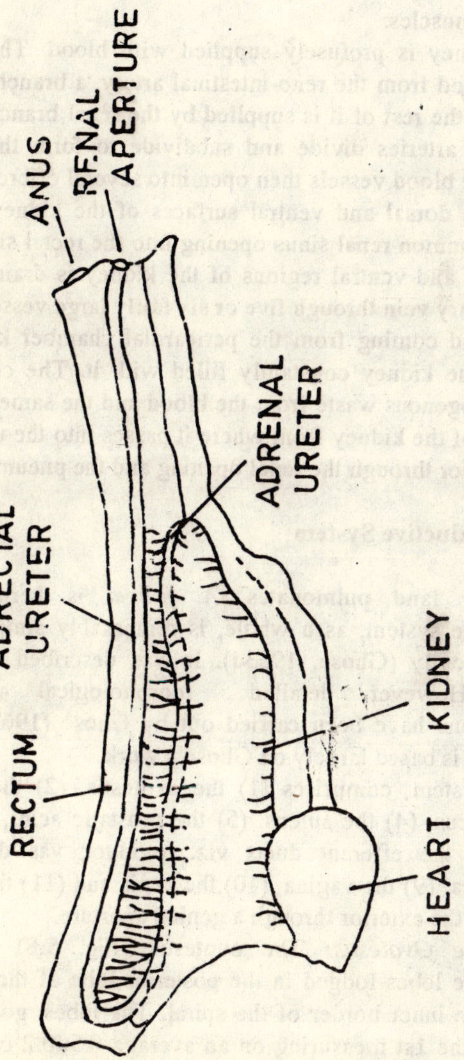

FIG. 5.7: Excretory organs of *A. fulica*

The wall of the adrenal ureter is thrown into folds and is supported by a thin basement membrane and a narrow band of longitudinal muscles. The wall of the adrectal ureter is supported by relatively thick and strong muscles.

The kidney is profusely supplied with blood. The posterior part receives blood from the reno-intestinal artery, a branch of the cephalic aorta while the rest of it is supplied by the renal branches of the rectal artery. The arteries divide and subdivide to form the afferent renal vessels. The blood vessels then open into several efferent renal sinuses both on the dorsal and ventral surfaces of the kidney. These join to form the common renal sinus opening into the rectal sinus. Blood from the anterior and ventral regions of the kidney is drained directly into the pulmonary vein through five or six fairly large vessels.

The fluid coming from the pericardial chamber keeps the central lumen in the kidney constantly filled with it. The cells of the folds absorb nitrogenous waste from the blood and the same is discharged in the lumen of the kidney from where it passes into the ureter and thence to the exterior through the renal opening and the pneumostome.

The Reproductive System

Like other land pulmonates, *A. fulica,* is hermaphrodite. The reproductive system, as a whole, is remarkably simple yet functions most efficiently (Ghose, 1963d). It was described earlier by Mead (1950a). However, detailed morphological and histological investigations have been carried out by Ghose (1963d). The present description is based largely on Ghose's work.

This system, comprises (1) the ovotestis, (2) the ovotestis duct, (3) the uterus, (4) the sulcus, (5) the prostatic acini, (6) the albumen gland, (7) the efferent ducts viz., oviduct, vas deferens, (8) the spermatheca, (9) the vagina, (10) the penis and (11) the genital atrium, opening to the exterior through a genital aperture.

(1) *The Ovotestis:* The ovotestis (Fig. 5.8) consists of 4-5 fingers–like lobes lodged in the posterior lobe of the digestive gland, towards the inner border of the spiral. The lobes, go on increasing in size from the 1st measuring on an average 0.5×0.2 cm to the last one measuring 0.8×0.4 cm. Each lobe of the ovotestis is divided into numerous minute lobules which are closely-packed together.

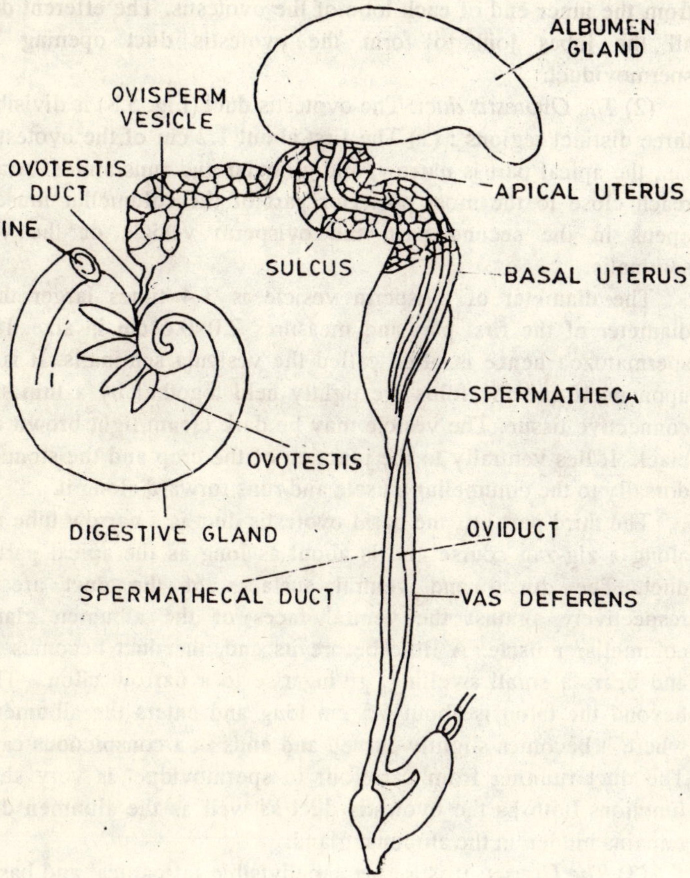

FIG. 5.8: Reproductive organs of *A. fulica*

Histologically, lobule consists of a thin capsule made of a single layer of flattened cells and round germs cells are attached to the capsule. In the centre of each lobule is a large lumen packed with sperms in most cases. Most of these produce only spermatozoa while a few produce both spermatozoa and ova. (Fig. 5.8). The spermatogonia are smaller than ovogoina so can be differentiated easily. Developing ova are enclosed in thin capsules and usually more than two ova are seldom observed in a lobule whereas the spermatozoa are numerous. Lumens of several lobules join together to form a narrow efferent duct; numerous such efferent ducts unite to form a larger duct and several such larger ducts finally join together and emerge as one single duct from the inner end of each lobe of the ovotestis. The efferent ducts of all the lobes join to form the ovotestis duct opening in the spermoviduct.

(2) *The Ovotestis duct:* The ovotestis duct (Fig. 5.8) is divisible into three distinct regions : (1) The first about 1.2 cm of the ovotestis duct i.e., the apical part is narrow, and globular and runs almost straight to reach close to the most posterior part of the columellar muscle and opens in the second part, the ovisperm vesicle or the vesicula seminalis.

The diameter of ovisperm vesicle is 3-4 times larger than the diameter of the first part and measures 2.0×0.6 cm in size. It stores spermatozoa hence is often called the vesicula seminalis. It is folded upon itself and the folds are tightly held together by a thin sheet of connective tissue. The vesicle may be dark cream/light brown or even black. It lies ventrally to the junction of the crop and the stomach and dorsally to the columellar muscle and runs forward along it.

The third part i.e., the basal ovotestis duct is a narrow tube running along a zig-zag course and is about as long as the apical part of the duct. The dorsal and ventral surfaces of the duct are placed respectively against the ventral faces of the albumen gland and columellar muscle. A little before its end, the duct becomes straight and bears a small swelling, giving rise to a narrow talon. The duct beyond the talon is about 0.5 cm long and enters the albumen gland where it becomes slightly curved and ends in a conspicuous carrefour. The duct running from carrefour to spermoviduct is very short and functions both as the ovotestis duct as well as the albumen duct and remains hidden in the albumen gland.

(3) *The Uterus:* It is a long sac divisible into apical and basal parts and is lodged in the posterior haemocoelic chamber, forming a tight jacket over the uterus.

The apical part (Fig. 5.8) is creamy white, large and much folded and measures about 2.5×0.6 cm. It is in the form of a loop and the inner concave margin of its basal end lies dorsal to the columellar muscle. Its wall bears numerous transverse folds.

The basal part of the uterus (Fig. 5.8) is dull brown, elongated, widest in the middle and narrow at both the ends. The basal end is narrower than the apical end. Towards the end of apical part, for a short distance, the wall of the uterus has transverse folds; along the rest of its entire length, it bears a large number of narrow longitudinal folds.

(4) *The Sulcus:* The sulcus (Fig. 5.8) or the sperm groove is a narrow, incomplete passage running along the inner concave margin of the spermoviduct. The sulcus can be separated from the uterus by two large folds of the wall of the spermoviduct — a broad, short and straight fold and a narrow, long, curved, hook-like fold. The first fold arises from the inner margin and the second from the lateral margin of spermoviduct. When the lateral fold is pushed outwards, it comes to lie against the opposite wall and the sulcus becomes a completely separate sperm channel and also acts as a barrier against the entry of eggs in the sulcus.

Ducts of the prostatic acini discharge their secretions in the sulcus close to the base of the inner fold.

(5) *The Prostatic acini:* Prostatic acini are creamy white, more or less conical structures resembling maize grains and arranged in several rows along the inner concave margin of the spermoviduct. Each acinus is made up of a large number of alveoli, opening into a central prostratic canal. These are closely packed together and are connected with the sulcus by their narrow ends, the broader end being directed outwards. The maximum number of acini recorded by Ghose (1963 d) exceeded 110. During the breeding season the acini increase in size but after oviposition they shrink considerably.

(6) *The Albumen gland:* This is a creamy-white, large smooth-surfaced and slightly curved gland (Fig. 5.8) made up of a large number of vesicles which open into the ciliated albumen canal. The latter leaves the gland as a short albumen duct which ends in the carrefour. Its ventral side is flat and the dorsal convex. It is round both at the apex and the base, but the base is broader than the apex. It lies anterolaterally and more or less parallel to the proximal part of the rectum and is placed between the basal ovotestis duct and the ovisperm vesicle ventrally and the digestive gland dorsally. During the breeding season, the gland becomes about 3 times larger than its usual size. The largest-sized albumen gland recorded by Ghose (1963d) measures 6.2×1.8 cm. After contributing albumen to the eggs, the gland again shrinks.

The gland consists of a large number of small, spherical or oval elongated vesicles, held together by very narrow and delicate strands of tissue. A narrow duct lined by gland cells is formed by the union of the lumens of several adjacent vesicles and opens in the albumen canal in the middle. The albumen canal is lined with an epithelium of medium-sized ciliated cells and it communicates with carrefour through a very short and narrow albumen duct.

Albumen is stored temporarily in the albumen canal from where it is driven to the apical uterus by the movement of cilia present both in the albumen canal and duct and in the common exit duct.

(7) *The Oviduct and the vas deferens:* The Oviduct (Fig. 5.8) is a narrowed anterior continuation of the basal uterus. The vas deferens (Fig. 5.8) is a continuation of sulcus and bears a small rounded swelling at its base, pressed firmly against the few most anterior prostatic acini. Both are narrow tubes with muscular walls and are enclosed by muscular tissues near their origin.

(8) *The Spermatheca:* It is an elongated, oval, thin-walled sac (Fig. 5.8) with a narrow duct. It is attached to the anterior end of the basal uterus with connective tissue.

(9) *The Vagina:* Anteriorly, the spermatheca joins the oviduct and forms the vagina. The vagina is a muscular tube and anteriorly it has a spherical swelling which divides it into (a) a long, narrow, apical vagina and (b) a short, broad, basal vagina, ending in the genital atrium.

(10) *The Penis:* The penis is a slightly curved and muscular structure. It is broad in the middle but narrows slightly at both the ends. The concave face is placed on the right and the convex on the left side. The preputial region is narrower but a constriction between the preputial and vergic region is not always well marked. At the junction of the two is a small chamber which opens in the genital atrium through a narrow passage in the penial prepuce.

(11) *The genital atrium:* It is a small, stout, muscular structure with a narrow lumen which serves as the common terminal passage for both male and female ducts. The lumen of the genital atrium opens to the exterior by a narrow genital aperture situated at the centre of a round whitish area called the genital area.

6

ECONOMIC STATUS

Evaluation as a Pest

A. fulica, the giant African snail is a pest mainly of the horticultural crops all over the world. As stated earlier it is a pest of vegetables, fruits, ornamentals and some plantation crops.

In India, it has been reported as a pest from Assam, West Bengal, Nagaland, Bihar, Orissa, Tamil Nadu, Kerala and Maharashtra. In addition, it is a menace on several islands of Andaman and Nicobar group, especially to vegetable cultivation and to zinnia, marigold, coffee seedlings and seedlings and fruits of papaya. It has also been observed causing damage to arecanut seedlings and fruit buds (Fig. 6.1), the apical meristem of rubber plants and its buds and contaminating and drinking rubber latex. A recent visit (September 1989) to Port Blair has given a clear-cut indication that although the snail continues to cause damage to vegetation, its number has been reduced considerably as a result of continuous efforts by the Indian Agricultural Research Institute, New Delhi and its Port Blair Substation (now merged with the Central Agricultural Research Institute, Port Blair) from 1964 to 1980, in collaboration and with the cooperation of the Department of Agriculture, Andaman and Nicobar Islands, Port Blair. Even if the Department of Agriculture, Andaman and Nicobar Islands, Port Blair and the Central Agricultural Research Institute, Port Blair follow the recommendations of Integrated Control Measures, this pest can be contained. However, it requires regular vigil as in the case of insect pests.

Workers on the giant African snail have reported the food/host plants of this pest. The largest list of 227 host plants has been given by Olson (unpublished) from Hawaii. His list includes, besides his own observations, observations of 20 other workers. The list of Olson has plants of 181 genera; out of these, 3 genera belong to Pterydophyta,

FIG. 6.1: *Achatina* crawls up on the stem of arecanut quite high

TABLE 6.1: List of some economically important host/food plants of *A. fulica*

Names of some host/food plants and their economic importance	Mead (1961)	Olson (1973) (unpublished)	Raut and Ghose (1984)
1	2	3	4
A. FRUITS			
Family : Caricaceae			
1. Papaya (*Carica papaya*) 1 A quick-growing small tree; starts fruiting in about a year; grown in most parts of India mainly in Assam, West Bengal, Bihar, Orissa, Uttar Pradesh, Gujarat, Kerala and Andaman and Nicobar Islands. Ripe fruits are relished; unripe ones are cooked as vegetable or pickled. Papain-a proteolytic enzyme and item of great commercial importance is obtained from the latex of immature green fruits. Besides being a proteolytic enzyme, papain is also used in the preparation of chewing gum, in textile and beer industry. It also forms an important ingredient of digestive medicines (Singh *et al.*, 1983).	Listed as a preferred host plant. Seedlings are completely eaten up; leaves, fruits, buds and flowers of older plants damaged. Occasionally bark also is eaten up.	Listed as a preferred host plant.	Seedlings are completely eaten away, feed 'greedily' on leaves; floral buds, flowers and fruits are also eaten/damaged; bark also damaged.

TABLE 6.1 (*Contd.*)

1		2	3	4
Family : Moraceae				
2. Bread fruit (*Artocarpus altilis*) A small fruit tree grown in India in Maharashtra and along the eastern and western coasts of southern region of the country. It is also grown on the Andaman and Nicobar Islands. Fruits cooked as vegetable; white, viscous gum obtained from the bark (Singh *et al.*, 1983).	2	A preferred host plant in some areas; green, overripe, rotting, even the dried fruit is "real favourite of the giant snail".	Listed as a preferred food plant.	—
3. Jack (Jak) fruit (*Artocarpus heterophyllus* Syn. *A. integrifolia*) A tree native of India and grown all over the country including the Andaman and Nicobar Islands. Fruits and seeds are relished as vegetable and also pickled; bark used for tanning hides.	3	3-4′ (90—120 cm) high seedlings may be chewed up overnight; bark, even of mature trees may be consumed; 'fallen fruit is a great favourite' of *Achatina*.	Listed as a preferred host plant.	—
Family : Musaceae				
4. Banana (*Musa paradisiaca* Syn. *M. sapientum*) A native of India; now grown extensively in Assam, West Bengal, Bihar Madhya Pradesh, Gujarat, Andhra Pradesh, Karnataka, Tamil Nadu, Kerala and Maharashtra. Unripe fruits are used as vegetable and also	4	Snails frequently found on leaves and trunks. Occasionally young banana fruits are rasped on account of which, when they mature, they fetch less price even though the edible portion remains unaffected. If some bananas	split open on the trees, they are consumed by the snails. Listed as a food plant of *Achatina* but not referred to as 'preferred'.	Young snails (up to 60 days old) do not attack the plants. As referred to by Mead (1961), Raut and Ghose also refer to the rasping of the young banana fruits on account of which they get 'badly scarred' and fetch less price in the market.

TABLE 6.1 (*Contd.*)

1	2	3	4
used in diarrhoea and dysentery; ripe fruits are relished by all; also useful in diabetes and several other diseases (Singh *et al.*, 1983). In South India, leaves are commonly used as plates for lunch, dinner and even in marriage parties.	Leaves also are reported to be damaged.		
B. VEGETABLES			
Family : Amaranthaceae			
5. Amaranth (*Amaranthus* spp.) Cultivated throughout India. Leaves and young stems are used as vegetable. *Amaranthus blitum* and *A. tricolor* are the two main species; *A. tricolor* is a rich source of Vitamin A; contains 2,500 to 11,000 International units per 100g.	5 A preferred food plant. Older plants are seriously attacked but younger ones are completely eaten up.	Listed as a preferred host plant i.e., most severely attacked by *Achatina*.	A preferred food plant all over India. Seedlings are favourite food of the snail of all age groups.
Family : Basellaceae			
6. Indian spinach (*Basella alba*) A common, twining succulent under-shrub grown for its leaves and tender stems, used as vegetable; fruits fleshy reported to be used as a dye of vegetable origin; may be used for dying jellies and sweets without any adverse effect on human health.	6 —	—	A preferred host plant of *Achatina*. Although the entire plant may be consumed by the snail but leaves are preferred.

TABLE 6.1 (*Contd.*)

1		2	3	4
Family : Compositae				
7. Lettuce (*Lactuca sativa* Syn. *L. scariola* var. *sativa*) 7a. *L. indica* *L. sativa* is a leafy herb cultivated throughout India; Leaves eaten raw; leaves yield lactucarium which has medicinal properties and used in case of bronchitis and asthma. *L. indica* is distributed in Sikkim, Meghalaya and Assam; leaves possess digestive properties.	7	Severe damage to *L. indica* reported; plants of all ages completely defoliated but seedlings are completely destroyed.	*L. sativa* has been listed as a preferred host plant.	*L. sativa* is a 'most preferred' host plant; young snails attack only the leaves, but older ones devour the whole plant.
Family : Cruciferae				
8. Broccoli (*Brassica oleracea* var. *italica*) It is grown only in some parts of the country.	8	—	Referred to as a preferred host plant.	—
9. Cabbage (*B. oleracea var, capitata*) Commonly grown practically all over India; stems short and thick, leaves packed into large head which is eaten as vegetable, both raw and cooked. A rich source of Vitamin A, 2000 International units/100 g; Vitamin B and riboflavin are 60, 30 μg/100g respectively.	9	Reports of severe damage to cabbage in Saipan. Mead reports Cabbage (*Brissica* spp.) and other cruciferous crops as the most preferred food plants in the Philippines.	Referred to as a preferred host plant.	A 'Most preferred' host plant. From seedlings onwards at all the stages of growth it is completely devoured. Behura (1955) also refers to cabbage as "one of the most preferred" food plants.

TABLE 6.1 (*Contd.*)

1	2	3	4
10. Cauliflower (*B. oleracea* var. *botrytis*) Grown practically all over India, especially in Uttar Pradesh, Karnataka, Punjab and West Bengal for its immature inflorescences which are eaten as vegetable and also pickled (Singh *et al.*, 1983).	10 This vegetable appears to have been covered under *Brassica* spp.	Referred to as a preferred host plant.	"Favourite food plant", Behura (1955) also reports *A. fulica* to be a "serious threat" to this crop in Mayurbhanj district of Orissa. Singh and Birat (1969) report that "serious-damage is inflicted to vegetables like cabbage and cauliflower so much so that sometimes transplantings have to be done two three times".
11. Radish (*Raphanus sativus*) Cultivated mainly in the Punjab, Uttar Pradesh, Maharashtra and Baroda. Roots, young tender leaves and fruits all are edible; eaten raw and also cooked as vegetable. Radish has medicinal value also and is given to patients suffering from liver/gall bladder problems. Juice of fresh leaves is used as diuretic and laxative (Singh *et al.*, 1983).	11 All parts of the plant reported to be attacked in Sumatra; damage is also reported from Saipan.	Listed as a host plant but not referred to as preferred.	Leaves and above ground portions of the stem are eaten.
12. **Family : Cucurbitaceae** A family of great economic importance especially regarding vegetables.	12 This family has been referred as "especially liable to damage".	This family as a whole has been listed as "preferred".	—

TABLE 6.1 (*Contd.*)

1	2	3	4
13. Bottle gourd (*Lagenaria siceraria* Syn. *Cucurbita sciceraria, C. lagenaria, Lagenaria vulgaris*) A climbing/trailing herb, cultivated throughout the country for its fruits which are used as vegetable said to be a good source of Vit. B and ascorbic acid (Vit.C) Seeds have medicinal value in dropsy. Dry, hard shells of fruits used as water bottle, bowl, blowing horn and musical instruments.	13 Seriously attacked in the Philippines.	Listed as a preferred host.	Bottlegourd not so much preferred as *C. maxima;* leaves are occasionally eaten.
14. Cucumber (*Cucumis sativus*) A trailing/climbing annual, cultivated for its edible fruits, eaten raw; contains 25.8 ug Zinc/gm on dry weight basis. Seeds also are edible, are diuretic and also yield edible oil.	14 While from some places there are reports of very little damage, from others serious damage is reported and often the entire plant is eaten up.	Listed as a preferred host.	"One of the most preferred food plants" and the damage is reported to be serious. Young plants are damaged beyond survival; older ones are defoliated, stem is often cut into pieces resulting in complete loss of the crop.
15. Melons a. Musk melon (*Cucumis melo*) b. Water melon (*Cirtrullus lanatus* Syn. *C. vulgaris, Cucurbita citrullus*) Musk melon is a creeping annual, cultivated throughout India particularly in the hot-dry north-western areas for its fruits which are eaten when mature; ripe seeds also are edible, cooling and	15 Both fruits and vines are preferred food plants. Young fruits are badly disfigured and become unfit for human consumption. In some cases, serious damage to the entire plant also is on record.	Musk melon and water melon both, have been referred as preferred hosts.	—

TABLE 6.1 (*Contd.*)

1	2	3	4
diuretic. Water melon is a climbing/trailing annual; cultivated throughout India for its cool and refreshing fruits; seeds also edible and yield oil (Sing *et al.*, 1983).			
16. Pumpkin (red) (*Cucurbita maxima*) Red pumpkin, a climbing/trailing annual; is grown for its fruits which are used as vegetable; contains 84 international units Vit. A, and 60 μg and 40 μg/100 g Vit. B_1 and Riboflavin respectively. Seeds yield resin having herbicidal properties (Singh *et al.*, 1983).	16 —	Listed a preferred host.	Seedlings attract snails of all age groups; leaves, twigs, floral buds and flowers are eaten up; stems chewed up, plants completely defoliated and stem is often cut into pieces.
17. Pumpkin (white) (*C. pepo*) White pumpkin also is a climbing/trailing hairy annual, commonly cultivated in North India for fruits, which are used mainly as vegetable (Singh *et al.*, 1983). Also used in preparing sweets.	17 Serious damage reported from different parts; leaves skeletonised and stems barked in Ceylon (Sri Lanka). In Chichi Jima seedlings had to be raised in protected boxes and then transplanted "to prevent them from being chewed off as they came up".	Listed as a preferred host.	—

TABLE 6.1 (*Contd.*)

1	2	3	4
18. a. Ridged gourd (*Luffa acutangula*) b. Smooth gourd (*L. aegyptiaca* Syn. *L. cylindrica*) *Luffa* spp. are trailing annuals, cultivated throughout India for their young tender fruits which are used as vegetable. Fibrous material obtained from ripe (mature) dry fruits used as bath sponge. Luffa sponges are also used as packing material and for stuffing pillows (Singh *et al.*, 1983).	18 *Luffa* spp. seriously attacked in the Philippines.	*L. acutangula* listed as a preferred host.	Both the species of *Luffa* are "favourite food plants" of the snail; plants are often defoliated and stem chewed up.
19. **Family : Leguminosae** This is a very important plant family as it contains pulse crops, a valuable source of vegetable protein for men and animals.	19 "Wide preference" for the plants of this family.	Family is listed as a preferred host.	—
20. Beans (*Lablab purpureus* Syn. *L. vulgaris*, *Dolichos lablab*) Cultivated all over the country, young pods used as vegetable. Also grown as cover crop and fodder (Singh *et al.*, 1983).	20 —	—	Leaves and twigs are eaten voraciously; sometimes fruits also are eaten up completely. As usual, preference is for the seedlings.
21. Cowpea (*Vigna unguiculata* Syn. *V. sinensis*) Cultivated all over the country; young pods used as vegetable; mature seeds cooked and consumed as a pulse.	21 Fruits, leaves and stems damaged; young plants completely destroyed, older ones skeletonized.	Listed as a preferred host.	—

TABLE 6.1 *(Contd.)*

1		2	3	4
22. Peas (*Pisum* sp.) *Pisum sativum* cultivated mainly in the Punjab, Himachal Pradesh, Haryana, Uttar Pradesh and Delhi as a favourite vegetable. Plants used for forage (Singh *et al.*, 1983); and pods, contains 139 I.U. Vit. A and 250 μg Vit. B_1/100g.	22	—	*Pisum* spp. listed as a preferred host.	—
Family : Malvaceae 23. Okra/Lady's finger (*Abelmoschus esculentus* Syn. *Hibiscus esculentus*) An erect herb cultivated throughout India mainly for its tender fruits which are used as vegetable. Ripe seeds contain about 20 per cent edible oil. Mucilage from stem and roots is used for cleansing sugar cane juice in the manufacture of *gur* (Singh *et al.* 1983). Fruits contain 58 I.U. Vit. A and 63, 60 μg/100 g Vit. B_1 and Riboflavin respectively.	23	Severe damage reported; plants practically defoliated and fruits damaged to such an extent that they become unfit for marketing.	Listed as a preferred host.	Sometimes plant is defoliated but flowers and fruits are 'food of choice'; young plants are completely eaten up.

TABLE 6.1 (*Contd.*)

1	2	3	4
Family : Moringaceae			
24. Drum stick (*Moringa oleifera* Syn. *M. pterygosperma*) A small tree, native of India; green fruits are used as vegetable; oil obtained from seeds used as medicine in gout and rheumatism. Wood pulp suitable for newsprint (Singh *et al.*, 1983). Contains 184 I.U. Vit. A and 50 and 65 μg/100g Vit. B_1, Riboflavin respectively.	24 Observed attacking this plant in the Philippines.	Listed as a host plant but not referred to as preferred.	Only the leaves of this plant are eaten.
Family : Solanaceae			
25. Brinjal/egg plant (*Solanum melongena*) An erect plant native of India. Grown all over India for fruits used as vegetable; fruits contain, 45 and 90 μg/100g. Vit. B_1 and Riboflavin respectively.	25 There are reports of damage to this plant from Ceylon, Saipan and the Philippines. In Ceylon bark of the stem is reported to be eaten away.	Listed as a host plant but not referred to as preferred.	—
26. Chillies (*Capsicum annum*) This herb is now cultivated throughout India; an important condiment crop grown mainly in Andhra Pradesh, Maharashtra, Karnataka, Tamil Nadu, Madhya Pradesh, Bihar and Rajasthan (Singh *et al.*, 1983). Raw, green and ripe red fruits used as spice and	26 Fruits, leaves and epidermis of stem are reported to be attacked in Sumatra.	*Capsicum* spp. is referred to as a preferred food plant.	Leaves and skin are eaten. Srivastava (1966) has also reported chillies as a food plant of *Achatina*.

TABLE 6.1 (*Contd.*)

1	2	3	4
condiment; green fruits also are pickled. Has medicinal value also. Green ones contain 454 I.U. Vit. A and 180 μg/100g Riboflavin respectively. Mature dry ones contain 576 I.U. Vit. A/100 g.			
27. Tomato (*Lycopersicon lycopersicum* Syn: *Lycopersicum Esculentum, Solanum Lycopersicum*) A common herb cultivated for its fruits in the entire country. Ripe ones eaten raw or cooked as vegetable; green immature ones also eaten as cooked vegetable. Red ripe ones also used to prepare soup, sauce and ketchup, hence it is cultivated on commercial scale.	27 There are reports of serious damage to foliage in Guam following the first heavy shower of the season but in Ceylon tomato is reported to be not damaged by *Achatina*.	Listed as a host plant but not referred to as preferred.	Extensive damage in nurseries but the crop is attacked when no other preferred food plants are available.
Family : Umbelliferae 28. Carrot (*Daucus carota*) An annual or biennial herb cultivated extensively in the Punjab, Haryana, Uttar Pradesh and Madhya Pradesh for its fleshy tap roots; eaten raw or boiled; carrot tops used as fodder; seeds useful in diseases of kidney and in dropsy. Carrot (root) is a rich source of Vit. A; 2000-4300 I.U. of Vit. A/100 g present.	28 There are reports of severe damage to both the "tops and the tuberous roots".	Listed as food plant but not referred to as preferred.	

TABLE 6.1 (*Contd.*)

1	2	3	4
C. ORNAMENTALS			
Family : Apocynaceae			
29. Oleander (*Nerium indicum* Syn. *N. odorum; N. oleander*) A shrub grown in gardens throughout India for its fragrant flavour; also grown as hedge plant. Root applied externally as a paste against ring worm; dried leaves are effective cardiac stimulant.	29 —	—	Leaves are eaten, rarely flowers also eaten.
Family : Araceae			
30. Money Plant (*Scindapsus* (*Pothos*) *aureus*) A climbing/trailing herb, cultivated as an ornamental practically all over the country.	30 —	—	This plant is one of the main sources of food of the snails in gardens, damage is rather serious.
Family : Compositae			
31. Cosmos (*Cosmos* spp.) *Cosmos* spp. is an annual herb cultivated all over India in gardens as an ornamental; one of the most common garden flower plants.	31 Reported damaged in the Philippines.	Listed as a food plant but not referred to as preferred.	A choice plant for snails of all age groups; twigs preferred; plants defoliated; epidermis eaten away; stem cut into pieces, damage serious.

TABLE 6.1 (*Contd.*)

1	2	3	4
32. Chrysnathemum (*Chrysanthemum* spp.) *C. carinatum,* an annual herb with different colour combinations; grown as an ornamental; *C. cinerariifolium,* grown in Kashmir and Nilgiris for getting the insecticide pyrethrum; highest pyrethrin contents present in the inflorescence when in full bloom; used as insecticide for mosquito coils for burning and for fly sprays and aerosols. *C. coccineum* cultivated in Assam; dried flowers do yield pyrethrum which is less toxic than that obtained from *C. cinerariifolium.*	32 —	—	Snails have a definite choice for the leaves and flowers.
33. a. French marigold (*Tagetes patula*) An ornamental grown in the gardens for its reddish yellow/orange flower heads. Flowers yield an essential oil used in perfumery; roots and seeds used as purgative.	33a —	—	One of the 'most preferred' plants for snails for all age groups. The older snails consume the entire plant; the younger ones avoid the hard portions. Leaves are 'avidly eaten' by the 20-25 day old snails.
b. African marigold (*T. erecta*) A bigger-sized marigold, flowers, leaves and stems all source of an essential oil; "an infusion of the plant used against coid, bronchitis and rheumatism".	33b —	*T. erecta* is in the list of host plants but is not referred to as preferred.	—

TABLE 6.1 (*Contd.*)

1	2	3	4
34. Sunflower (*Helianthus annuus*) An ornamental herb; seeds are source of vegetable oil which is used for cooking, as lubricant, for burning and in the manufacture of soap, paint and varnish; oil cake used as a cattle feed.	34 —	—	Snails climb up to the top of the plant and consume the floral buds and flowers.
35. Zinnia (*Zinnia angustifolia* Syn. *Z. linearis*) A herb grown in gardens all over the country, seedlings grown in autumn, flowering in winter.	35 Damage recorded from the Philippines to *Z. angustifolia*.	Like the other two flowers *Cosmos* and *Tagetes* of family Compositae, *Zinnia* spp. also is not referred as preferred although it is listed as a host plant.	*Z. angustifolia* seedlings are 'one of the most favoured' food plants; occasionally grown up plants also are attacked.
Family: Crassulaceae			
36. Air plant (*Bryophyllum pinnatum*) It is a succulent ornamental, cultivated in the tropics, grown successfully in drought conditions also. Little plantlets develop on the edges of leaves. Flowers are borne in considerable numbers at the top of stems. It is a medicinal plant and plant extract is applied on boils.	36 Densely-growing plants in uncultivated areas of Hawaiian islands are a huge stock of 'unlimited source' of preferred food for the snail.	Listed as a preferred food plant.	—
Family: Malvaceae			
37. China rose (*Hibiscus rosa sinensis*) An evergreen ornamental shrub, flowers yield dark-purplish dye used in preparing shoe polishes. Leaves have medicinal value. Hedges of this plant	37 Damage to flowers and leaves of *Hibiscus* spp. reported. Hedges of this plant observed loaded with *A. fulica*.	Listed as a preferred food plant.	A favourite host plant of the snail; although all parts of the plant are attacked, damage is caused more to flowers and leaves.

TABLE 6.1 (*Contd.*)

1	2	3	4
are quite common. Flower decoction used in bronchial problems; root-decoction recommended in veneral diseases.			
Family: Nyctaginaceae 38. Bougainvillea (*Bougainvillea* spp.) A climbing spiny shrub, grown in gardens for colourful petal-like bracts.	38 Both small and large snails attack this plant.	Listed as a host plant but not referred to as preferred.	Only leaves are eaten.
Family : Portulacaceae 39. Rose-moss (*Portulaca grandiflora*) A herb commonly grown as an ornamental.	39 —	—	Snails older than 30-40 days attack this ornamental severely. Succulent plant body is 'avidly eaten'. There are reports of complete destruction of this ornamental in gardens of Kerala, West Bengal, Assam and Bihar.
D. PLANTATION CROPS **Family : Euphorbiaceae** 40. Rubber (*Hevea brasiliensis*) A plant of great economic importance, presently being grown in South India and the Andaman and Nicobar Islands. Also grown in Assam, Tripura and Goa. This tree is the source of "best natural rubber obtained from its latex". Rubber is an important component of tyres, tubes, foot wear and water proof	40 In Ceylon damage to young rubber plants is serious. There are reports of bark being eaten away and plants getting killed. Leaves eaten and growing tips, young seedlings and buds are also damaged. They also drink latex, sometimes considerable amount, also contaminate it in the collection cups with their mucus and excreta etc.	Listed as a host plant put not referred to as preferred.	—

TABLE 6.1 (*Contd.*)

1	2	3	4
clothing. Also used in preparation of hose, belts, foam sponges, mats and various electrical goods. Latex adhesive and rubber thread are other important items of commercial importance (Singh *et al.* 1983).			—
Family : Palmae			
41. Arecanut (*Areca catechu*) An erect slender-stemmed palm grown in Karnataka, Kerala, Tamil Nadu, West Bengal, Assam, Maharashtra. It is also grown extensively on the Andaman and Nicobar Islands. Nuts used as a masticatory and are chewed along with the leaves of *Piper betle* Also used in veterinary practice as a vermifuge against tapeworms. Arecanut husk is used for the preparation of hard boards and brown wrapping paper. It is also used in medical sciences. Arecanut has a prominent place in the Indian religious observances and mythology.	41 —	—	Young snails are found in large numbers in the inflorescence, "slight scratches" were observed on the soft bark. In Andamans, the author has observed slight damage to leaves in nurseries. Snails also climb up to the top of *Arecanut* tree; cause damage to flowers and occasionally also to newly-formed nuts.

TABLE 6.1 *(Contd.)*

1	2	3	4
Family : Rubiaceae			
42. Coffee (*Coffea arabica*) An evergreen shrub, now commonly grown in the Nilgiris, Karnataka and Travancore; grown to some extent in Andamans also. Dried seeds are roasted grouped and brewed to prepare a stimulating beverage. Seeds of *C. canephora* mostly used in the manufacture of "instant coffee".	42 In the absence of other food plants, young leaves are reported to be attacked in Sumatra, slight damage to berries is reported from Malaya.	Listed as a host plant but not referred to as preferred.	Raut and Ghose nave not listed it as a host plant of *Achatina*, but the author and his co-workers have observed damage to coffee seedlings on South Andaman.
Family: Sterculiaceae			
43. Cocoa (*Theobroma cacao*) It is a small tree not indigenous to India but now cultivated in South India and Orissa. Seeds are a source of beverage and are also used in confectionery. Cocoa-butter obtained from seeds is used in the preparation of chocolates, cosmetics, perfumery and as a base for certain ointments. A by-product of Cocoa processing industry is used for extracting theobromine which possesses diuretic properties and acts as myocardial stimulant and smooth muscle relaxant also; shell is used as fuel (Singh *et al.*, 1983).	43 Mead has reported that in Ceylon the giant African snail was observed congregating and feeding on the flowers, broken pods, seeds, and on seedlings in the nurseries. Mead observed the damage to range from 'slight to severe'. There were also reports of two month old seedlings being readily attacked and killed. There are also reports of young plants being seriously attacked and blossoms being damaged.	Listed as a host plant but not referred to as a preferred one.	Not listed as a host plant.

TABLE 6.1 (*Contd.*)

1	2	3	4
E. TUBER CROPS			
Family : Convolvulaceae 44. Sweet potato (*Ipomea batatas*) Cultivated throughout India for sweet edible tubers; tubers are cooked and eaten; tubers also used commercially in the manufacture of alcohol, syrup, glucose and starch; starch used in the manufacture of textiles, paper, cosmetics etc. (Singh *et al.*, 1983).	44 Reports differ from place to place. The vine indeed provides a cover for a large number of snails but damage to leaves and exposed tubers is reported from some places, damage at some places is little, at other places it is just attacked or is the chief food plant	Has been referred to as a preferred host plant.	
Family : Diascoreaceae 45. Yam (*Dioscorea alata,* Syn. *D. purpurea, D. rubella*) A climbing shrub, now cultivated in Assam, West Bengal, Tamil Nadu, Madhya Pradesh and Vadodara (Gujarat). Tubers are used as vegetable. (Singh *et al.*, 1983).	45 Leaves are "favourite of the snail"; young plants eaten and serious damage is reported.	Referred as a preferred host.	
Family : Euphorbiaceae 46. Cassava/Tapioca (*Manihot esculenta*) A small shrub, now grown in Kerala, Tamil Nadu, Karnataka and Andaman and Nicobar Islands. Commercially starch, sago, semolina and flour are obtained from it (Singh *et al.*, 1983).	46 From some places there are reports of no damage, while at other places damage is reported to be serious; there are reports of young stems being heavily peeled off and killed, leaves also are damaged.	Listed as a host plant but not referred to as preferred.	—

TABLE 6.1 (*Contd.*)

1	2	3	4
F. OILSEEDS AND PULSES			
Family : Euphorbiaceae			
47. Castor (*Ricinus communis*) Cultivated chiefly in Andhra Pradesh, Maharashtra, Karnataka and Orissa (Singh *et al.*, 1983). Also cultivated in Delhi, Uttar Pradesh, Bihar for getting oil out of its seeds. Castor oil is used as lubricant and purgative, also used commercially in the preparation of phenyles, hair oils, hair fixers and lipsticks; castor oil cake is used as a fertilizer (Singh *et al.*, 1983).	47 —	—	**Young snails generally skeletonize the leaves; seedlings are 'completely' devoured by older ones. Damage is also caused to leaves, bark, twigs and fruit coat.**
Family : Leguminosae			
48. Soybean (*Glycine max*) Soybean is a bushy herb now cultivated as a pulse/oilseed crop mainly in the Punjab, Haryana, Delhi, Uttar Pradesh, Maharashtra, Gujarat, Naga Hills, Manipur and Kashmir (Singh *et al.*, 1983). A rich source of protein, (43.2 per cent); in addition contains 710 I.U. Vit. A and 730 and 760 µg/100 g Vit. B_1 and Riboflavin respectively. Yields an edible oil, oil used in different industries and in the household as a frying medium; soybean is also used in enzyme (lipoxidase)	48 —	***Glycine max*** **referred as a preferred host plant.**	**Snails eat leaves and twigs of *Glycine* spp. voraciously, sometime even the pods are eaten completely. Preference is shown for growing regions and seedlings by snails of all age groups; damage serious.**

TABLE 6.1 (*Contd.*)

1	2	3	4
preparation widely used in baking industry and for bleaching flour. The left over after oil extraction is used as cattle feed (Singh *et al.*, 1983)			
49. Groundnut/peanut (*Arachis hypogea*) An erect or trailing herb, now widely grown in South India, Gujarat, and Maharashtra for its pod; seeds very rich in protein (26.31 per cent) and Vitamins particularly B_1 (900 μg/100g) and Riboflavin (300 μg/100g); nuts roasted and eaten, peanut butter prepared from the nuts; oil used as a frying medium and in the manufacture of vegetable ghee (vanaspati); cakes used as a source of high-protein livestock fodder; groundnut shell is used as fuel and in the manufacture of coarse boards.	49 There are reports of young plants being completely destroyed, older ones reported defoliated. There are reports of leaves, stems and nuts being attacked.	Listed as a preferred host.	
50. a. Black gram (*Vigna mungo* Syn. *Phaseolus radiatus*) b. Green gram (*Vigna radiata* Syn. *Phaseolus* radiatus, *P. aureus*) Grown all over India as major pulse crops. Grown also for green manuring.	50 "*Phaseolus radiatus* seedlings totally destroyed", leaves of older plants skeletonized.	*Phaseolus aureus* (mung bean) listed as a preferred host plant.	

TABLE 6.1 (*Contd.*)

1	2	3	4
G. FIBRE CROPS			
Family : Malvaceae			
51. Cotton (*Gossypium* spp.) A shrub cultivated practically all over the country; staple cotton obtained from surface of seeds and used for the manufacture of textiles, hull (outer covering of seed from which staple cotton and cotton linters (short fibres arise) used as cattle feed and fertilizer; cotton seed oil is edible, also used in the manufacture of vegetable ghee (vanaspati) soaps, paints and varnishes, toilet preparations, in leather industry and as lubricant; cotton seed cake used as cattle feed and fertilizer.	51 In Mauritius in 1911 "It was responsible for a good deal of damage to cotton seedlings. . . ."	Listed as a host plant but not referred as preferred.	Young plants of *G. herbaceum* are completely destroyed; leaves, flowers, floral buds and fruits of older plants are attacked.
Family : Tiliaceae			
52. Jute (*Corchorus capsularis*) Economically a very important seasonal fibre crop grown mainly in west Bengal and Bihar; grown also in Uttar Pradesh, Orissa and Assam. Fibre obtained from stem is used for the manufacture of gunny bags, packing cloth, covers for cotton bales, ropes, carpets, coarse cloth and paper. Jute sticks give high yields of greaseproof paper required by confectionery	52 —	—	Young plants are seriously attacked and the leaves of grown up ones are eaten only by the young snails; Occasionally epidermis is also scrapped but the twigs are never eaten.

TABLE 6.1 (*Contd.*)

1	2	3	4
industry; jute is also mixed with wool for preparing cheap blankets (Singh *et al.*, 1983).			
H. FOREST TREES			
Family : Verbenaceae			
53. Teak (*Tectona grandis*) Grown in Western Peninsula, Central India and Bihar. Wood termiteproof, used for manufacture of ships, bridges, buildings, furniture; musical instruments, plywood and black board also prepared; wood waste used for preparing fibre board etc. (Singh *et al.*, 1983).	53 There are reports from Java of up to 90 per cent lethal damage to young plants while in Sumatra one year old plants are damaged.	Listed as a food plant but not referred to as "preferred".	—

TABLE 6.1 *(Contd.)*

1	2	3	4
Family : Mimosaceae			
54. Kokko/Siris (*Albizia lebbeck* Syn. *Mimosa lebbeck, Acacia lebbeck*) A tree suitable for avenues, wind breaks and shelter belts. Andaman is the main region of this tree. Also grown to a "limited" extent in Assam, West Bengal, Tamil Nadu, Madhya Pradesh, Punjab and Uttar Pradesh. Leaves and twigs used as fodder; grown for shade in coffee and tea plantations; wood excellent for high class furniture and interior decoration etc., wood pulp is used for the manufacture of writing paper and newsprint (Sing *et al.*, 1983).	54 A. *lebbeck* referred as preferred host in Saipan.	Referred as a preferred host.	

one genus is a Gymnosperm and 177 genera belong to Angiosperm. It is worth noting that out of 177 genera of Angiosperm, only 32 are monocotyledonous and the rest are dicotyledonous. Thus our major food crops, the cereals, are spared by this very destructive pest. Olson has referred to only 88 plants as "preferred host plants".

Mead (1961), in his monumental publication on *Achatina fulica* has given a list of 86 plants which are liked by this snail to varying degrees.

Raut and Ghose (1984) have given a list of 66 plants from India and the damage reported is from severe to negligible. They have stated that nearly 90 per cent plants cultivated in India are "acceptable" to *A. fulica.* A comparative account of economically more important food plants listed by Olson, Mead and Raut and Ghose especially those which are found in India also, has been given in Table 6.1. There are certain anomalies and certain similarities in the observations of these authors and those of the author and his co-workers. These have been discussed below:

Amongst fruits, papaya (Figs. 6.2 and 6.3), breadfruit and jackfruit are damaged severely. Olson refers to these as preferred hosts. In addition to seedlings and leaves, fruits also are eaten (Mead, 1961). Raut and Ghose have not listed breadfruit and jackfruit as host plants of this snail but they have listed papaya, whose seedlings are damaged and leaves, floral buds, flowers and fruits also are eaten. Banana finds place in all the three lists but Olson has not listed it as a preferred host plant. Mead as well as Raut and Ghose have referred to the rasping of young banana fruits, on account of which the scars, remain even after the rasped fruits mature. Consequently such bananas fetch less price in the market even though the edible portion remains unaffected. Mead refers to damage to leaves also.

Amongst vegetables, amaranth and lettuce have been mentioned as preferred in all the three lists. Mead states that older plants of amaranth are attacked seriously but the younger ones are completely eaten up. Raut and Ghose state that the seedlings are favourite food of snails of all age groups. Regarding lettuce, Mead has reported that seedlings are completely destroyed and the plants of all age groups are completely defoliated. Raut and Ghose have referred to this plant as a "most preferred" host. They state that young snails attack only the leaves but the older ones eat away the entire plants.

Vegetables of family cruciferae appear to be favourite of this snail. Olson refers to broccoli, as a preferred host but others have not listed

FIG. 6.2: Infants and young *Achatina* feeding on papaya leaves

FIG. 6.3: Giant African snail congregating in large numbers on the stem of papaya

it. However, cabbage and cauliflower find place in all the three lists. Olson refers to both these vegetables as preferred host plants. Mead reports severe damage to cabbage and other cruciferous crops. Raut and Ghose refer to cabbage as "most preferred" and cauliflower as "favourite food plant" of this snail. Radish has been listed by Olson but not as preferred, Mead has reported damage to all parts of the plant while Raut and Ghose report damage to leaves and "above ground" parts. It is not clear why Olson has not referred it as a preferred host. It is possible that his report on radish may be observation of some other worker or in the presence of some more preferred host, this snail may not have damaged radish.

Like Cruciferae, Cucurbitaceae and Leguminosae are two other families whose vegetable plants are on the whole preferred by the snails. While Mead refers to Cucurbitaceae as "especially liable to damage", Olson refers to this family as a "preferred one". Olson has listed bottlegourd, cucumber, musk melon, watermelon, pumpkin, ridgegourd and smoothgourd as preferred and by stating as preferred, Olson means "most severely attacked". As regards bittergourd, Olson has referred it as preferred; Mead has not mentioned bittergourd in particular as a food plant of *Achatina* and Raut and Ghose have referred to leaves as "casually eaten". The author and his co-workers have never observed any damage to bittergourd on the Andamans, where the snail is abundant. Mead refers to serious damage to bottlegourd but Raut and Ghose report that bottlegourd is not so much preferred as *Cucurbita maxima.* Mead has listed *C. pepo* and not *C. maxima* as a host plant of this snail and he has mentioned serious damage to it. *Cucumis sativus* (cucumber) finds place in all the three lists; Mead mentions the damage from very little to severe, Olson has listed it as a preferred host whereas Raut and Ghose consider it as "one of the most preferred" food plants of this snail. Raut and Ghose have not listed melons, but Mead refers to both fruits and vines of melons as preferred hosts. Olson also has listed melons a preferred hosts. According to Mead, *Luffa* spp. are seriously attacked. Olson has listed ridged gourd as preferred whereas Raut and Ghose have listed both ridged gourd and smooth gourd as favourites of this snail. The author also has observed heavy damage to *Luffa* spp. (Fig. 6.4).

Family Leguminosae as a whole has been referred to by Mead and Olson both as preferred. While Olson has referred to cowpea and peas as preferred, Mead has listed only cowpea; Raut and Ghose have listed only beans as the leguminous host of this snail.

FIG. 6.4: *Achatina* have almost completely defoliated a part of the vine of *Luffa*

Vegetables of families Malvaceae, Moringaceae, Solanaceae and Umbelliferae also have been reported as host plants of this snail. Some of these have been referred to as preferred while some others have been just listed. The author and his co-workers have observed severe damage to okra; at seedling stage, the seedling is eaten up completely. Olson has listed okra as a preferred host. Mead reports that plants are practically defoliated and the fruits are damaged to such an extent that they become unfit for marketing. Amongst the solanaceous plants, brinjal, chillies and tomato have been listed, but Raut and Ghose have not listed brinjal. Olson has listed brinjal but not as preferred host but chillies have been listed as preferred. Fruits, leaves and epidermis of the stem of chillies have been reported to be eaten by Mead; Raut and Ghose also have listed this plant as a host of the snail. The author also has observed damage to leaves and stem of this plant. Although tomato has not been listed by Olson as a preferred host, Mead reports severe damage to foliage, whereas Raut and Ghose have reported "extensive damage" in the nurseries.

Amongst the ornamentals, money plant has been listed only by Raut and Ghose. This ornamental is grown all over the country including the Andaman and Nicobar group of Islands. The author has observed the snail feeding upon this creeper in the laboratory as well as in the field. Ornamentals of family Compositae, like other plants of this group have been observed to be favourite of this snail at least in India. Raut and Ghose have reported damage to cosmos, chrysanthemum, marigold, sunflower and zinnia. Olson has listed cosmos, marigold and zinnia but none of these have been referred to as preferred. Mead has reported damage to cosmos and zinnia. It is really not understandable how Olson has not recorded cosmos, marigold and zinnia as preferred hosts. The author has observed especially marigold to be "a most preferred hot plant" of this snail on the Andaman islands and at Delhi. Once this snail gets this plant, it leaves everything else and finishes it up. On the author's suggestions, one gentleman at Port Blair saved his papaya seedlings by transplanting several thick rows of marigold plants around his papaya saplings and these grew up and started fruiting. Raut and Ghose (1984) have reported that due to the menace of giant African snail in 24-Parganas, Howrah, Midnapur and Calcutta of West Bengal and at Balasore in Orissa many cultivators have given up commercial cultivation of marigold. In the face of all these observations it is difficult to understand how Olson has failed to enlist it as a preferred host. The only reason appears to be that it is based on the observation of some other author as he has himself stated that his list is based (besides his own observations), on the observations of some other workers also.

Like marigold Olson has listed *Zinnia* as a host plant but not as a preferred one, while Mead has reported damage to *Zinnia* in the Philippines and Raut and Ghose have reported *Zinnia* seedling as "one of the most favoured food". The author also has observed damage to this plant on the Andamans. Mead (1961) has referred to *Bryophyllum* as a densely-growing plant in uncultivated areas of Hawaiian islands acting as an unlimited source of preferred food for the snail. Olson also has listed it as a preferred host plant. However, this ornamental has not been listed by Raut and Ghose nor have the author and his co-workers observed it as a food plant of *Achatina*. China rose has been listed as a preferred host plant of *Achatina* by Olson and Raut and Ghose. Mead has stated that hedges of *Hibiscus* spp. were loaded with giant African snail and all parts of the plant were attacked. *Bougainvillea* has been referred as a host plant by Mead, Olson and Raut and Ghose but none have referred it as a preferred host. Raut and Ghose have reported complete destruction of *Portulaca.* The author and his co-workers also have observed *Achatina* feeding on *Portulaca* in the laboratory.

Amongst the plantation crops, arecanut, rubber, coffee seedlings and cocoa are attacked by this snail. Olson has not listed any of these as preferred host and has not listed arecanut even as a host plant of this snail; rubber, coffee and cocoa have not been listed by Raut and Ghose. Mead has reported serious damage to young rubber plants in Ceylon. He has also reported young seedlings, buds and growing points getting damaged by this snails. There are also reports of seedlings getting killed. The snail also drinks latex and contaminates it. Mead has observed damage to cocoa ranging from slight to severe.

Oilseed crops are important commercial crops. Castor, soy bean and groundnut have been listed as host plants of *Achatina.* Castor has been listed only by Raut and Ghose who report that seedlings are completely damaged by the snail. Soy bean has been referred to as a preferred host by Olson. Raut and Ghose have reported that the leaves and twigs also are eaten voraciously. Groundnut has not been listed by Raut and Ghose, but Olson has listed it as a preferred host. Mead has reported complete destruction of young plants and defoliation of older ones.

Like several other leguminous crops, green gram has been listed by Olson as a preferred host. Mead has reported complete destruction of the seedlings of *P. mungo.*

Mead has reported damage to leaves of young plants of tobacco. In addition, bark of the plant at the base is eaten away with the result that the stem may break up easily when the wind blows. Even otherwise such plants either die or have poor growth, probably because there is disruption in the upward movement of food material from the roots.

Amongst the fibre crops, jute and cotton have been listed as host plants of *Achatina.* Cotton finds place in all the three lists. Although Olson has not listed it as preferred, Mead reports good deal of damage to cotton seedlings in Mauritius. Raut and Ghose have reported complete destruction of young plants in addition to attack on leaves, flowers, floral buds and fruits. Jute has been listed only by Raut and Ghose obviously because it is economically a very important crop grown in India mainly in West Bengal and Bihar. They have recorded serious damage to young plants and damage to leaves and epidermis of the twigs.

Amongst the forest trees, damage to only two trees teak and

kokko/*Albizia* have been reported by Mead and Olson. Raut and Ghose appear to have restricted their observations only to field crops. Mead and Olson both have referred *Albizia* as a "preferred host". Olson has not listed teak as a preferred host but Mead has reported up to 90 per cent damage to young plants in Java; in Sumatra, one year old plants are damaged.

According to the published records and unpublished reports maize (*Zea mays*) is the only cereal reported to have been slightly damaged or nibbled occasionally by this snail. Although maize finds place in all the three lists, but the damage caused appears to be only occasional. The author has not seen any damage to maize crop on the main island of South Andaman although maize is grown in those areas where snails are abundant. On the main land also, the author neither observed any damage to maize nor received any report of damage to this crop.

In general the snail prefers dicotyledonous plants; amongst these, a large number of its preferred food plants belong to the families Cruciferae, Cucurbitaceae, Compositae, Malvaceae and Leguminosae, whether these be vegetables, ornamentals, pulses, oilseeds or belong to any other group of plants.

In spite of this snail being such a destructive pest, there are plants which it does spare or occasionally causes negligible damage. Such plants can be named as non-host/ non-food plants of *Achatina* (Table 6.2). Some of these have been listed by Mead, Olson and Raut and Ghose, but none of these have been listed as preferred except bittergourd by Olson. These have been tabulated in Table 6.2. In the last column of this table observations of the author and his co-workers have been given. These observations related mainly to the Andaman and Nicobar islands as well as observations in the laboratory and some other places on the mainland i.e., Pusa (Bihar), Guwahati (Assam), Baripada (Orissa), Madras (Tamil Nadu), Calcutta (West Bengal) and Delhi. This table will be discussed in detail in the chapter dealing with the management of this snail.

Extent of Damage: Although lot of information on the qualitative evaluation of damage to plants by the giant African snail is available, very little is known about the quantitative aspect. Raut (1982) has done exhaustive study of the quantitative aspect of damage caused to 8 species of economically important plants belonging to six plant families i.e., Leguminosae, Cucurbitaceae, Compositae, Malvaceae, Caricaceae and Basellaceae. The plants included beans, sponge gourd, cucumber, gourd (*C. maxima*) marigold, okra, papaya and the Indian spinach (*Basella alba*). He carried out experiments in the garden in field cages in order to give them fair amount of natural environment. He observed that on the whole young snails preferred young plants, while older ones attacked young as well as grown up plants. The author also has observed large number of young snails congregating on one single leaf of plant and consuming it soon. Raut (1982) observed that young bean plants were consumed completely leaving no trace of the plant and stems of many were cut into pieces. Serious damage was caused to gourd, sponge gourd and cucumber. Some plants were completely defoliated. Twigs, stems, leaves, floral buds and young fruits were eaten voraciously. Out of 139 okra plants, floral buds of 83 were damaged so badly that no fruits could be obtained from these. Marigold plants, as expected, were completely defoliated and denuded. The snails were observed "to drink latex from the papaya fruits" on account of which those fruits did not remain marketable. The amount of consumption varied with 'plant species' as well as with the rate of infestation by the snails of different size-groups.

Deka *et al.*, (1985) have studied the damage caused by *A. fulica* to plant species of 11 families including Cucurbitaceae, Malvaceae and Solanaceae. They have observed maximum damage to cucurbitaceous plants. This family has been enlisted also by Mead, Olson and Raut and Ghose. The author also has observed serious damage to plants of this family (except bittergourd). Although Deka *et al.*, (1985) consider Malvaceae and Solanaceae as tolerant to *A. fulica*, Raut (1982) has reported serious damage to okra (Malvaceae). Okra and cotton (Malvaceae) find place in the lists of Mead, Olson and Raut and Ghose. China rose has been listed by Olson and Raut and Ghose. Similarly amongst the solanaceous plants, brinjal has been listed as a host plant of this snail by Mead and Olson and *Capsicum* sp. finds place in all the three lists. The author also has observed serious damage to okra, brinjal and *Capsicum* sp. on the Andamans. Just as family Cucurbitaceae is a family of preferred host plants of *Achatina*, yet bittergourd is not eaten but occasionally just nibbled, in the same way it is possible that there may be some malvaceous and solanaceous plants which this snail may not be feeding upon or causing only minor damage. However, on the basis of the vast amount of information available, even if one or two malvaceous and solanaceous plant species are not damaged, it will be more appropriate to call these families as susceptible rather than tolerant to the snail.

TABLE 6.2: List of some economically important non-host/non-food plants of *A. fulica*

Names of the non-host/non food plants and their Economic importance	Mead (1961)	Olson (1973) Unpublished	Raut and Ghose (1984)	Observations of the author and co-workers
1	2	3	4	5
A. VEGETABLES AND CONDIMENTS **Family : Alliaceae** 1. Onion (*Allium cepa*) It is a biennial herb grown practically all over the country, mainly in Maharashtra, Tamil Nadu, Karnataka, Andhra Pradesh, Madhya Pradesh, Punjab, Haryana and Uttar Pradesh. Immature/mature bulbs are consumed raw or eaten as cooked vegetable; bulbs are consumed as stimulants and in dysentry etc. Onion juice is reported to increase diuresis; its active principle is glycollic acid (Singh *et al.*, 1983). Now due to	Mead (1961) has quoted Weel (1948, 1949) as having observed no damage to *Allium* spp.	1. Not listed as a host plant.	Not listed as a host plant.	We have not observed any damage to onion or garlic neither in the field nor in the laboratory.

TABLE 6.2 (*Contd.*)

1	2	3	4	5
introduction of new varieties of onion is being grown both as Kharif as well as Rabi crop and is being exported in large quantities.				
2. Garlic (*Allium sativum*) Garlic is a *Rabi* crop and is grown practically all over the country, wherever climate suitable for its cultivation is available. It is grown on a very large scale in Gujarat. Bulbs are used as condiment and for flavour; used as carminative and gastric stimulant. Presence of allicin in garlic inhibits the growth of several micro-organisms, hence its antibacterial activity; therefore, it is widely used in several infectious diseases. Garlic juice is used as ear drop and in colic and flatulence. Garlic taken internally destroys worms and is used for external application in skin diseases	2. —	—	—	

TABLE 6.2 (*Contd.*)

1	2	3	4	5
(Singh *et al.*, 1983). Garlic has been found to have curative effect on rheumatism. Now-a-days a deodorized preparation of garlic known as garlic pearl is available in the market.				
Family : Cucurbitaceae 3. Bitter gourd (*Momordica charantia*) It is a trailing or climbing herb; it is grown all over the country for use as cooked vegetable. Bitter gourd is reported to be useful in diabetic cases.	Not listed as a host plant.	3. Listed as a preferred host plant.	Leaves of this climber "are casually eaten".	We have not observed damage to this vegetable crop. We have observed bitter gourd being grown freely on the Andamans. The statement of Raut and Ghose (1984) that leaves are "casually eaten" means that damage, if any is negligible.
B. ORNAMENTALS **Family : Balsaminaceae** 4. Garden Balsam (*Impatiens balsamina*) It is an ornamental herb with beautiful flowers, grown throughout the country including the Andamans during the rainy season. Its seeds are edible. Alcoholic extract of flowers is reported	Not listed as a host plant.	4. Not listed as a host plant.	Leaves and flowers are attacked, but twigs are rarely eaten; however, the snail shows preference for fallen leaves and flowers.	In Andamans, where giant African snail is a serious threat to horticultural crops we have seen lines after lines of garden balsam. It is the common flower plant grown on these islands. The other very common flower plant, which people like to

TABLE 6.2 (*Contd.*)

1	2	3	4	5
to have antibiotic activity against some pathogenic fungi and bacteria (Singh *et al.*, 1983)				grow, is marigold which is a favourite food plant of the snail and is consumed in no time unless provided proper protection regularly.
Family : Cupressaceae 5. Oriental vitae (*Thuja orientalis* Syn. *Biota orientalis*) It is a perennial shrub with deep green coloured leafy crown, grown all over the country as an ornamental; it makes avenue in parks and gardens. It can be grown as a potted plant also. Stems and leaves yield tannin.	Not listed as a host plant.	5. Listed as a host plant, but not as a preferred one.	Listed as one of the many ornamental plants as a host plants of *A. fulica,* but damage to this ornamental plant has not been referred to as serious.	The author has observed the giant African snail resting on *Thuja* in Orissa and Andaman, but did not observe any damage.
C. PLANTATION CROPS **Family : Palmae** 6. Coconut (*Cocos nucifera*) In India coconut is grown on the Andaman and Nicobar group of islands, also in the states of Maharashtra, Tamil Nadu Kerala and West Bengal. Fibrous parts of the fruit are used for preparing mats, carpets, ropes etc., the	No damage from any where reported, although the snails were observed swarming on fronds of coconuts.	6. Coconut does find a place in Olson's list but has not been referred to as preferred. Olson has mentioned that he has listed "plants fed on in varying degrees by *A. fulica*". May be he may have got some reports of very slight damage.	Not listed as a host plant.	During our about, sixteen years experience, on a number of islands of Andaman and Nicobar group, we did not observe any damage neither at the nursery stage nor to the grown up plants although snails were very frequently observed congregating and resting on coconut trees.

TABLE 6.2 (*Contd.*)

1	2	3	4	5
hard shell used as bottle of 'hookas'; the half shells used as bowls/ cups. The kernel is eaten raw and also dried as "Copra" for eating and for extracting oil which is used as hair oil, as cooking medium, in the manufacture of soap etc. Water in the cavity of the green unripe fruit is relished as drink.				mainly on their trunks (Fig. 7.1)
Family : Theaceae 7. Tea (*Camellia sinensis* Syn. *C. thea, C. theifera, Thea sinensis*) A very important economic plant of great commercial importance for local consumption as well as for export. It is an evergreen shrub, but is generally prunned down to low bush. In India, cultivated mainly in Assam; Darjeeling and Jalpaiguri districts of West Bengal, Travancore, the Nilgiris, Malabar, Dehradun	There are conflicting reports. No damage has been reported from Ceylon, but from other places there are some reports of young shoots, flowers and young leaves being attacked.	7. Listed as a host plant but not as a preferred one.	No serious damage has been observed; in the the event of non-availability of other food plants, only leaves in some gardens of Jalpaiguri were observed damaged. However, in the nursery, seedlings are attacked.	So far there are no reports from Assam where tea cultivation is done on a large scale. During the author's visit, no damage to plantation was observed and no report of damage to tea plantation was received.

TABLE 6.2 (*Contd.*)

1	2	3	4	5
and Kumaon. Leaves are cured and used as beverage. Tannins of tea waste are used in plywood industry for preparing hot setting plywood adhesive (Singh *et al.*, 1983; Raut and Ghose, 1984). A good foreign exchange earner.				
D. OTHERS **Family : Gramineae** 8. Rice (*Oryza sativa*) Rice is the most important cereal eaten by maximum number of people in the world. In India, although it is grown all over the country, it is the main cereal crop in Assam, West Bengal, Bihar, Orissa, Eastern Uttar Pradesh, south Indian states and Andaman and Nicobar group of Islands. Rice straw is used for making straw boards paper pulp and mats. Rice straw is also used for feeding cattle, mainly where	*Achatina* does not feed on the plant but loss to the crop may be caused since they remain sticking to the plant and on account of their weight the plant may break.	8. Not listed as a host plant.	Not listed as a host plant.	The author and his colleagues never observed *Achatina* on the paddy crop; however, in some vacant paddy fields, after the crop had been harvested, sometimes, some snails were observed hiding during day in the big cracks in the field.

TABLE 6.2 (*Contd.*)

1	2	3	4	5
wheat is not grown/cannot be grown at all. Rice bran oil is used in the preparation of soaps and cosmetics. Rice bran is used as fuel, soil conditioner, bedding material for animals and poultry and packing material. It also forms a part of animal feed.				
9. Sugarcane (*Saccharum officinarum*) It is a tall, perennial plant grown mainly in Bihar, Uttar Pradesh, Punjab and Haryana; stems are chewed as such; cane syrup is taken as such and used for preparing jaggery (*gur*), sugar and vinegar. By-products of sugar industry are used for various purposes e.g., in the preparation of rum, industrial alcohol and synthetic rubber etc. Bagasse is used as a fuel in	They are reported often congregating on leaves in large numbers; generally, no damage was reported; there were some reports of damage to leaves and leaf axils. However, the damage was of no real significance.	9. Listed as a food plant of *Achatina* but has not referred it as a 'preferred host'.	Not listed as a host plant.	Sugarcane is grown in some areas in Andamans, but neither we observed any damage to the crop nor the villagers ever reported to us any damage to their sugarcane crops.

TABLE 6.2 *(Contd.)*

1	2	3	4	5
the preparation of *gur* sometimes a mixture of bagasse and *gur* is used as cattle fodder. Bagasse is also used for manufacturing bagpaper, wrapping, writing and printing papers and card boards. The importance of the crop is further evident from the fact that in India there are two ICAR Institutes carrying on researches on various aspects of sugarcane only.				
Family : Bromeliaceae 10. Pine apple (*Ananas comosus*) It is a biennial herb now cultivated in Tamil Nadu, Kerala, Karnataka, Andhra Pradesh (coastal areas), West Bengal, Assam, Tripura, Orissa, and Andaman and Nicobar group of islands. Fruits are eaten; leaves are a source of strong silky fibre used for preparing strong cords,	Neither there is any record of snails feeding on pine apple nor any damage has been observed or reported from Hawaiian Islands or Ceylon.	10. Listed as a host plant but not referred, as a preferred one.	Not listed as a host plant.	Pine apple is grown abundantly on the islands of Andaman and Nicobar group but we did not observe any damage anywhere. The author did not observe any damage to the crop in Assam, West Bengal, Orissa, Bihar and Tamil Nadu — all these states being snail-infested areas.

TABLE 6.2 (*Contd.*)

1	2	3	4	5
fishing nets etc. Pine apple is also preserved and canned. Pine apple waste is a source of oxalic acid.				
Family : Pandanaceae 11. Screw pine (*Pandanus odoratissimus*) It is a shrub grown in the coastal region and on the Andamans. A very popular perfume is obtained from its spadices. This perfume has been in use in India since ancient times for scenting clothes, bouquets, lotions, cosmetics, soaps, hair oils, tobacco and *agarbatti*. Lower part of the fruit is eaten in south India. Leaves are used for making baskets and paper etc. Ladies use sweet smelling spathes as hair ornaments.	There are reports of no damage to *Pandanus* spp. but in some cases it was occasionally fed upon.	11. Listed as a host plant but not as a preferred one.	Not listed as a host plant.	On the Andamans, *Achatina* and Screw pine, both are found but the snail was never observed feeding upon it although they may rest there.

TABLE 6.2 (*Contd.*)

1	2	3	4	5
Family : Piperaceae 12. Betel pepper (*Piper betle*) It is a climbing shrub which is cultivated for its leaves mainly in Uttar Pradesh, Bihar and Andaman and Nicobar group of islands. It is very much relished as masticatory, along with lime, betel nut and katechu. Many like to use tobacco also along with betel pepper.	No damage was reported from any where.	12. Not listed as a host plant.	Not listed as a host plant.	The author did not observe any damage on the Andamans.
Family : Labiatae 13. Holy basil — *Tulsi* (*Ocimum sanctum*) It is a medicinal plant grown all over India, in houses, gardens and temples. Oil from leaves has anti-bacterial properties; leaf decoction is useful in cases of common cold. Aqueous leaf extract is said to be useful in cases of bronchitis and is also applied locally in cases of skin diseases (Singh et al., 1983).	Not listed as a host plant.	13. Not listed as a host plant.	Not listed as a host plant.	On Andamans, *O. sanctum* is widely grown in houses and temples. Their growth is very luxuriant and no damage by *A. fulica* or any other land snail has been observed by the author and his co-workers.

Evaluation as a feed for Fowls and Pigs and as a Part of Human Diet

Snail meal prepared from dehydrated *A. fulica* is a nutritive diet. The dehydrated flesh of this snail contains the amino acids lysine and arginine respectively 1.33 and 2.33 times more than dehydrated whole chicken egg (Mead and Kemmerer, 1953). It is also a rich source of protein as the dehydrated foot and the visceral mass both contain about 61 per cent protein (Table 6.3).

It is evident that flesh of *Achatina* is a very nutritive material. Mead (1982) very rightly states that it is really anomalous that there are certain regions which have a store house of protein in the form of this snail, yet people in those regions are suffering from malnutrition. If *Achatina* could be used as a supplementary diet of fowls, domestic animals and human beings, it "would coincidentally provide a positive pest control measure and exploit an impressive biomass heretofore lost to the environment" (Mead, 1982). Kekauoha (1966 b) has observed heavy population of *Achatina* in Hawaii, the average being 7.75 snails/m^2. Abbas and Gautam (1975) have recorded, on an average, the maximum population of the 102 snails/m^2 in the aestivating pockets of Makkapahar on South Andaman. If the snails be small to medium-sized, weight of the snails may be taken on an average about 32g per snail, the biomass of the snail on Hawaiian and South Andaman islands works out to be 2.4 and 33 metric tonnes/ha respectively (Mead, 1982). As is well known, major portion of the diet of *Achatina* comprises dead and decaying animal and plant tissues; till recently very little use has been made of this "machine" (*A. fulica*) which converts the scavenging material into highly nutritive flesh (Mead, 1982).

TABLE 6.3: Chemical analysis of dehydrated flesh of *A. fulica* (After Mead, 1961)

(in percentage)

Chemical constituents	Foot	Visceral mass
Protein	60.95	60.99
Calcium	0.25	0.26
Potassium	2.75	3.35
Phosphate (P_2O_5)	11.78	35.34
Phosphorus	3.67	10.70

There is a lot of scope for making use of this rich source of nutrition. Under favourable climatic conditions and availability of plenty of food material, this snail population can produce huge amount of biomass within a few months. In China and other snail-infested Indo-Pacific regions live or immediately killed unprocessed snails have been utilised for feeding fowls and pigs. In Indonesia during the rainy season 25 kg of snails and during the dry season the same weight of snails are sold for 1 and 3 American dollars respectively (Mead, 1982). On South Andaman, the author has seen some foreign breeds of pigs maintained by the army being fed on *Achatina*. At Calicut (South Andaman) the author observed a pig keeper, trapping *Achatina* during nights by putting on the pathways of the snail, bisected pieces of already fruited papaya stems and collecting these two-three times in buckets for feeding to his pigs (Srivastava *et al.* 1971). This snail was fed to the local breed of pigs of Andaman *Sus andamanensis.* Each pig ate, on an average 260 snails weighing about 3.8 kg per day. In Sarawak, the small-sized giant African snails were fed to soft-shelled fresh water turtles meant for the market (Mead, 1982).

At the centre for Animal Research and Development Bogor, Indonesia "the most" vigorous programme of processing the flesh of giant African snail for animal feed is going on (Creswell and Kompiang, 1979; Kompiang, 1979 a, b). The process of preparing snail meal and "snail silage" both have been developed and tested on young chickens. If the snails were boiled for 15 minutes before sun drying, it produced a more effective snail meal (without shell). Such snail meal has 60 per cent protein (or dry weight basis) and is nutritionally at par with fish meal. Addition to snail meal only up to 15 per cent in the diet led to proportionate increase in weight (Kompiang, 1979 a, b).

"Snail silage" is prepared by mixing snail's flesh with corn (1:1) or cassava (2:1) and is kept in closed containers for anaerobic fermentation. Although the nutritional value of silage is lower than that of snail meal, considering the abundance of the snail during the rainy season, when sun drying is practically not possible, it is much easier to make snail silage. However, preparing snail meal/silage is a laborious job and may not be feasible on commercial scale. For preparing snail meal/silage labour has to be employed for collecting, cleaning and finally marketing. Naturally all this will involve expenditure and the snail meal/silage thus prepared may not be cheap. The farmers, can, nevertheless process the snail's flesh for their own poultry/animals. While doing so they will be reducing the damage by the snail to their kitchen gardens and vegetable gardens.

As regards the possibility of imparting bad taste/flavour to the eggs and animals' flesh, Weel (1948) has found "no ill flavour" in the eggs of the fowls fed on *Achatina*, but Kondo (1952) states that on Guam people "believe" that the flesh of the pigs fed on giant African snail acquires an "undesirable taste". However simply "believing" may be different from "actual experience".

Mead (1982) states that since 1949 continuous efforts have been made to encourage the use of *A. fulica* as a part of human and domestic animal diet. According to Vosburgh, (1950) "The aborigines of interior Taiwan" readily cook and consume *Achatina*. The author observed Taiwanese fishermen lodged in jail at Port Blair cooking and eating *Achatina* gladly. After collecting the snails from the jail compound, they broke the shell and separated the foot from the rest of the body by a sharp knife. In this way, when the number of feet they needed for cooking were collected, they rubbed with dry wood ash to remove slime; later washed with water. The cleaned and well-washed feet were bisected, cut into convenient small-size and fried in the frying pan without fat for a minute; then a little common salt was added and further fried for 2 to 3 minutes. During this process, the flesh becomes softer. The fried (without fat) flesh was washed with water. Then quite a good amount of onion and chillies were fried in fat and the without fat-fried pieces of flesh of *Achatina* were added to it, fried for about 2 to 3 minutes and sufficient quantity of *Ocimum sanctum* leaves were added for flavour, removed from the frying pan and eaten. It is advisable to starve the snails overnight before processing for the table (Srivastava *et al.*, 1973). The author himself along with a number of senior officers of the Department of Agriculture, Port Blair tasted the snail preparation publicly at Port Blair to prove that there was nothing harmful in *Achatina;* however, the residents of the islands could not be persuaded to switch over to this rich source of protein. Their plea, was that when they did not eat it during the Japanese occupation when there was acute shortage of food material, why should they eat it now?

Persistent efforts have been made at commercial level to use snails for human consumption. In Mauritius, a project was started to encourage the consumption of the snail *Helix aspersa* available in plenty "in the humid areas of its central plateau", but it was not found to be commercially viable. There does not appear to be any possibility of *Achatina* being accepted for human consumption in Mauritius. In the Indo-Pacific region, on the Seychelles, in 1980, a restaurant had been serving *A. fulica* to the tourists and residents as "Seychelles delicacy". Since this country has plenty of fish available for consumption, whether under such circumstances the human population of this island will ever shift to *A. fulica* as a source of protein, is doubtful (Mead, 1982).

While the consumption of land snails by human beings in the Indo-Pacific region has been very limited, it is being eaten by the humans for centuries in Europe, Africa and Middle East. In the late seventies, Turkey, Yugoslavia and Romania exported 40,000 tonnes of land snails to France (Mead, 1982). In recent years, demand for land snails has been increasing in the USA and Canada and hundreds of tons of snails are being imported every year (Mead, 1982).

In the mid-seventies Taiwan started exporting frozen 50-60 mm-sized *A. fulica.* In 1975, 200 tonnes of canned frozen small-sized *Achatina* were exported to France; in 1977, 1500 tonnes of *Achatina* were exported to Europe and North America, out of which 90 per cent was exported to France which cost US $ 3,000,000. Export of *Achatina* is likely to increase substantially in the near future (Mead, 1982).

Frozen achatinas have been exported from the Hainan island of China to France since mid-seventies. A businessman of Java has been exporting the "soft parts" to France. The snails for export are collected by the local people; this is a source of additional income to them. Thailand also is exporting achatinas. There is demand for snails from abroad for canning. A businessman in Thailand exports the snail after treating it with salt and boiling to remove mucus. Export of Achatinas from Thailand has resulted in the reduction of its numbers.

However, some exporters started despatching "achatinas" in the guise of French "Escargot Achatine" and since there is difference in taste, to do away with this deception the French Government has passed regulations under which in addition to labelling as *Achatina* the country of origin also has to be mentioned. The U.S.A. also has now certain regulations regarding the import of snails.

Achatina continues to be exported to Europe, U.S.A. and Canada. Efforts are being made in Ghana and elsewhere in West Africa to culture the large-sized species of *Achatina* on commercial scale mainly for the local consumption.

In India, about 20 years back, efforts were made to export live *Achatina* of a particular shell-size from the Andamans to France because then it was extremely abundant on those islands; but it could not materialise. Now efforts are being made again in this direction. The

Central Inland Capture Fisheries Research Institute, Barrackpore (West Bengal) — an institute of the Indian Council of Agricultural Research, New Delhi, has been trying to rear this snail for export. Now, they claim to have developed a technology for the collection and processing of *Achatina fulica*. They also claim to have disseminated the technology to the exporters and thus have encouraged them to take up trade in edible molluscs. They also state that Marine Products Export Development Authority (MPEDA) "is getting orders from other countries for the processed snail meat" (ICAR Annual Report, 1988-89). There was a project on this snail with the Indian Agricultural Research Institute, New Delhi with a Research Station at Port Blair for about 16 years. The breeding technology is already known and collection from the field does not need any technology. Work on *Achatina* has been done for many years and is still in progress at the College of Science and Technology, University of Calcutta. These two institutions could have been of help to the Institute at Barrackpore. Atleast Calcutta was so near that the Scientists could have gone and learnt. As regards processing technique, probably the Barrackpore Institute is not aware that the processing technique also has not to be developed by research. The snail is already being processed and canned in Indonesia, Thailand and Taiwan for export. Some scientists can be sent to one of these places for a short-term training.

Let us hope that the processing and export of *Achatina* picks up in India and snails are collected from the field also for this purpose. This, in addition to earning foreign exchange, will also reduce the number of snails and, in the words of Mead (1982) would "coincidentally provide a positive pest control measure and exploit an impressive biomass heretoforth lost to the environment".

7

MANAGEMENT

The snail's nocturnal habit, its strong protective shell, its capacity to repair the damage caused to the shell, capacity to secrete mucus and the capacity to undergo hibernation/aestivation to face the unfavourable climatic conditions, stand in the way of control of this pest. It is a prolific breeder and since it is hermaphrodite, every snail is capable of laying eggs and in large numbers. Thus, the reduction in population due to the adoption of control measures is soon compensated by those left behind by laying large number of eggs. Thus, for a successful control of this pest, its habit, habitat and ecological aspects have to be taken into consideration. It is also necessary to bear two important facts in mind. Firstly, that this snail is not only a crop pest but also a scavenger as 74.8 per cent of its diet consists of scavenger type food (Olson, 1973). Thus, its role in nature as a scavenger cannot be ignored. Although control measures have to be adopted against it but the aim should be reduction of its population below the economic injury level and not eradication, so that the balance of nature is not upset. However, it is not advisable to allow its spreading to newer areas, which are so far free from this menace. Secondly, as far as possible, efforts should be made to adopt such control measures which may not pose the problem of hazard and the problem of pollution of environment.

There are several ways and means for the management of this pest and each method has its own merits and demerits. These methods are being discussed in the following pages:

Collection and Destruction of Snails and the Egg Masses

This is the easiest and simplest method which neither requires any sophisticated equipment nor very high skill. Any one adopting this method of control can pick up the method of approach within a few days. Of course undoubtedly it does require the proper guidance of an expert. Peterson (1957b) also has referred to reduction in population of this snail on Guam by collection and destruction. Olson (1973) states that baiting and hand collecting both are complementary to each other. If the collection is done before dawn and after dusk, large number of snails can be collected. If the snails are collected even for two hours during the night, preferably around mid-night, the number collected will be more than that collected throughout the day. Egg collection also is not a difficult job, as it can be done even during day time as snails lay eggs on the moist soil surface or just a little below it. The number of eggs collected and destroyed means not allowing those to hatch and thus, reducing the number of snails which could have been menace to the crops.

Although collection and destruction of the snail and eggs is easy, yet it is expensive as it involves a large number of man-hours and man-days. If these man-days/man-hours were utilised in doing some other work one could earn more.

Destruction of eggs poses no problem. The eggs collected should be crushed and disposed off. As regards disposing off the collected snails there are different views. Some people are of the view that in the sea shore areas, the collected snails can be disposed off in the seas. However, if this is to be done, the snails should be thrown in the sea far away from the shore so that none may be able to come back to the shore and crawl away on land. In the inland areas, where there is no sea, this cannot be done. Some people suggest to dig a pit and bury, but soon it starts emitting foul smell. Therefore, it is advisable to bury it at a depth of about 70-80 cm and in the fields away from the populated areas. If the snails are buried in cultivable fields, these act as organic manure as the snail's flesh contains Nitrogen, Potassium and Phosphates also. Mead (1961) has referred to the use of snail as a fertilizer in Chichi Jima and the reason for better crops in Chichi Jima at least partly is the use of snail as fertilizer and in Guam also where the "liquor" is drained off and diluted with 10 parts of water before adding to soil (Mead, 1961). On the Andamans certain cultivators stated that they got high yield of brinjal and cucurbits by the use of the snail as manure/fertilizer. To standardise the use of *Achatina* as a fertilizer, Gupta *et al.,* (1980) used snail's flesh at the rate of 20 kg/4 sqm plots and recorded significant increase in the production of a local variety of chillies in two locations, Haddo and Bloomsdale of South Andaman. If the snails are buried with shell, there is the possibility of soil turning alkaline. In Ceylon (Sri Lanka), soils of some tea growing areas turned alkaline because large number of giant African snail were buried with shells in those fields, with the result that those fields became unfit for growing tea (Mead, 1961). Thus, snails should be buried either without shell or at any place large number of snails with shell should not be buried.

Snail collection and egg collection can be arranged in the form of campaign either by voluntary agencies or by the government. At the government level, school children can be utilised for collecting snails and their egg masses on Sundays and holidays either early in the morning or late in the evening. As already stated earlier, egg collection can be done at any time during day. There may be separate egg collecting and snail collecting squads. If arranged at government level, schools can be closed for a few days and students utilised for this purpose. However, disposal of the snails and egg masses, must, in any case by done only under the personal supervision of some senior official in his/her presence. This is very important and any deviation from this is likely to do more harm than good. There is no dirth of manpower in India and it can be usefully utilized in the snail collection campaign.

Incentive to the snail collectors: Some incentive in the form of cash award and certificate may be given to the snail collectors, fixing some minimum weight limit of the snail, which may qualify them for the award/certificate. More than two decades back when the snail menace on the Andamans was at its peak, the Andaman administration had started the practice of making payment against the numbers (not weight). Later the practice had to be abandoned.

Destruction of Hiding Places of the Snail

During the Season of Snail's Activity

During the season of snail's activity, it has been observed that removal or even thinning of the hedges around gardens, reduces the snail menace. Peterson (1957b) goes a step further. He suggests that "if the garden areas of farm plots were surrounded by a belt of cleared land, a minimum of 20 feet (6 metres) wide, in which absolutely no plants are allowed to grow, invasion of crop areas usually could he held to a minimum". If this 6 metre wide fallow field is frequently ploughed, it will be all the more difficult for the snail to crawl up to the cropped field.

During the Season of Inactivity

When the snails undergo hibernation/aestivation, they remain hiding in inactive state under moist shady places in loose soil. Such hibernating/aestivating pockets should be located and destroyed along with the hibernating/aestivating snails. By doing so, on the return of the favourable climatic conditions, only those snails will be able to revive their activity who happen to have remained alive in some small pockets here and there. Since the number of such snails will be small, the loss caused by these snails to the cultivator will be much less unless these snails are allowed to remain alive and active to increase their numbers by prolific breeding. Since during hibernation/aestivation, larger numbers can be located in smaller areas, it is easier to collect and kill and it would involve less manpower and naturally less financial outlay (Raut, 1978).

In South Andaman, location and destruction of aestivating pockets along with the aestivating snails was commenced in 1973 and was continued regularly for three years (1973, 1974, 1975) in 11 villages. Population of snails in the aestivating pockets showed a decline year after year because of the regular destruction of aestivating snails (Table 4.11). It was also observed that the population of aestivating pockets in Makkapahar which is far away from the town (20 km away) had the highest population per square metre in all the three years and the villages nearer to the town i.e., Dolliganj, Nayagaon, Shadipur, Schoolline and Pahargaon, which were hardly 5-8 km away from Port Blair, had comparatively much lower aestivating snail population per square metre probably because people in these 5 villages were more enlightened and carried out control of the snail regularly (Abbas and Gautam, 1975; Gupta *et al.*, 1978). This is the first published record of location and destruction of aestivating pockets of *A. fulica* as a part of control measure. Raut (1978) also has suggested that location and destruction of aestivating pockets should form a part of control programme.

Growing Non-food Plants

In the regions where giant African snail may be causing serious damage, effort should be made to change the cropping pattern and grow those plants (Table 6.2) which the snail does not feed upon. While doing so, the climatic conditions of the infested regions as well as the economic importance of the crop will have to be taken into consideration. The snail does not cause damage to coconut (Fig. 7.1), pineapple and tea, but these crops cannot be grown every where although all the three are economically important. Thus, coconut thrives in the sea shore climatic regions, pineapple grows in sea shore climatic regions as well as in the regions having heavy rains like Andaman and Nicobar islands, and certain regions of West Bengal and Assam. There are tea growing areas in the south and in parts of West Bengal and Assam. Tea is not grown on the Andamans as not only climate but soil also matters. Betle pepper is another commercial crop and wherever the climate permits it can be grown. It is widely grown in many parts of the country including parts of Uttar Pradesh, Bihar, West Bengal and the Andamans. Sugarcane is another cash crop which the author has himself seen being grown in some snail-infested locations of South Andaman. Amongst the vegetables, onion and garlic are not attacked by the snail and can be grown in snail-infested locations of Bihar and Assam as *Rabi* crops. Onion is *Kharif* crop also can be grown in snail-infested locations even though rainy season is the season of activity of the snail also. Cucurbitaceae is a family, whose plants, on the whole, are preferred food plants of *Achatina* as is evident from Table 6.1. Bitter gourd also belongs to this family and is grown practically all over the country excluding probably the hilly tracts. Mead (1961) has listed it as a host of *Achatina* but has not referred to it as a favoured food plant. Raut and Ghose (1983a) have referred to the twigs and leaves of this creeper being damaged, but the stem, bark, floral buds, flowers and fruits have been referred as "not consumed". While listing this plant as a host of *Achatina* in their publication of 1984, they have stated that leaves are "causally eaten", i.e., the damage, if at all, is negligible especially because the floral buds, flowers and fruits are "not consumed". The author during his work on *Achatina* for over one and a half decades never observed any damage to this crop by *Achatina* in the snail-infested Andamans. It is not clear how Olson (unpublished) has listed it as a favoured food plant of *Achatina*. However, on the basis of the observations of Mead, Raut and Ghose and of the author, bitter gourd can be safely recommended for growing in the snail-infested locations.

Another cucurbitaceous vegetable is bottle gourd, which is damaged by *Achatina* but is comparatively much less preferred, than *Cucurbita maxima* and many other cucurbitaceous vegetables (Raut and Ghose, 1984). If the seedlings are protected, the creeper will grow

FIG. 7.1: The snail resting on coconut stem — causing no damage

well and there will be lot of fruiting. As regards the ornamentals, *Thuja orientalis* and the garden balsam can be grown safely in the snail-infested locations. The author has seen snails resting on *Thuja* at Port Blair and at Baripada (Orissa) without causing any damage. The snail has a special liking for marigold, but we have seen rows and rows of balsam growing and flowering all over South Andaman. As regards cereals, wherever possible, paddy cultivation can be taken up. Damage to this crop has neither been observed by the author and his co-workers nor there is any published or unpublished report of damage to this crop by *Achatina*. Maize has been listed by Olson as a host — not a preferred one and Raut and Ghose have mentioned that seedlings are "often attacked". Mead also has mentioned damage to young seedlings otherwise only "slight damage" is caused to leaves. The author and his co-workers have never observed any damage to maize crop by the snail on the Andamans but have not checked the seedling stage of maize. It can thus be concluded that if the maize crop is protected at the seedling stage, maize cultivation can be taken up in snail-infested locations also. There is no report of damage to wheat crop also by the snail. Thus cereals e.g., wheat, paddy and maize can be cultivated safely in the snail-infested locations but maize requires a bit of care at the seedling stage.

Holy basil (*Ocimum sanctum*) is a medicinal plant and does not find place in any of the three lists. It is grown all over the country including Andamans and other snail-infested locations. This plant is not damaged at all by the snails. Researches are going on, on the medicinal properties of this plant at the All India Institute of Medical Sciences, New Delhi.

Growing Trap Crops

If a preferred food plant of the snail has to be grown in a snail-infested location due to one reason or another, it can be protected by growing a trap crop all around the preferred food crop so that the snail may get trapped in the crop grown all round and may not reach the main crop. Since olfactory sensitivity does not appear to be much developed in the snail, it would remain busy in feeding on the outer rows of plants even if on the inner side a more preferred food plant is growing.

As already mentioned earlier in Chapter 6, on the author's advice papaya seedlings could be given protection from the snail at Port Blair by transplanting marigold plants all around, papaya grew up and started fruiting in about a year.

Raut and Ghose (1983b), as a result of their field-cage experiments, have suggested the growing of "non-crop plants" all around the economically important crop plants like soybean (*Glycine max*), lettuce, cabbage, beans, tomato (*Lycopersicon* sp.), pumpkin (*Cucurbita maxima*), okra, papaya, castor and cotton. They tried four non-crop plants out of which, as in Andamans, marigold provided maximum protection. However, marigold itself has now become a commercial crop in India and is being grown in many villages, which have facilities to transport its flowers to nearby towns. Thus, use of marigold as a trap crop or as a non-crop plant to protect crop plants does not appear to be feasible in all the situations and it can be decided according to local situation. However, the use of *Bryophyllum pinnatum* as a non-crop plant all around the main crop as used by Raut and Ghose (1983b) appears to be feasible in all the situations. This plant can be used as a trap crop. Mead (1961) has referred to this plant as "providing almost unlimited source of preferred food" in the uncultivated areas of the Hawaiian Islands.

Deka *et al.,* (1985) have given an altogether different suggestion. Their field investigations have indicated that plantating of the repellent malvaceous and solanaceous crops along the boundary, at least in two rows, and susceptible plants grown in the middle was found to reduce the damage completely. The observations and recommendations of Deka *et al.,* do not agree with the observations and recommendations of Raut and Ghose (1983 a) and the author.

Biological Control

On account of the gradual increase in the environmental pollution, these days there is a trend all over the world to discourage the use of pesticides for pest management even though we have not yet reached a stage when we can do away with the pesticides. Efficacy of the biological control agents in causing mortality of the snail and techniques for rearing these on large scale for releasing in the snail-infested locations have to be developed. For the biological control of *A. fulica,* a number of biological control agents, i.e., predators, parasites and disease-causing organisms are on record. However, biological control in relation to Economic Malacology is still in an undeveloped state and Mead (1979) considers it to be in the "alpha" stage. It is evident from the fact that so far, there is no record of the population of any pest snail having been contained entirely by the biological control agents to the extent that they may cease to cause economic damage to plants. To make the biological approach to the snail problem a real success, there is need for continuous research to test their efficacy in the laboratory as well as in the field and for developing mass rearing techniques of biological control agents in the laboratory.

Predators

In general, predators are non-specific. They may prefer certain living beings but generally do feed upon others also. Thus predator of a particular species of land snail may feed upon some other land snails as well as other organisms also. If land snail pests are predated upon by non-snail predators, it would be better as these can be used along with molluscicides also. Under such circumstances the molluscicide, while killing the pest snail would spare the non-snail predator. The predator should be such that its utility may be possible in all snail-infested regions, whether the location be a cultivated area, uncultivated area, sea shore region like Tamil Nadu, Kerala and Andamans or inland area like Bihar and Assam. The predator should be able to establish itself after the release. The biotic potential of the predator should be higher or atleast more or less the same as that of the pest. Finally, the predator, even if it may be an effective one, may not play a destructive role towards any crop. The author is not aware of any predator of any land snail having all or even most of these qualities. The predators of *Achatina* hitherto on record are discussed in the following pages.

The predators may be invertebrate or vertebrate.

(A) *Invertebrate Predators*

Invertebrate predators of *A. fulica* can be divided into three groups (i) Flatworms (*Turbellaria*) (ii) Arthropoda and (iii) Predatory snails.

(i) *Turbellaria* (Planaria) : An endemic nocturnal, black, 7-striped turbellarian flatworm *Geoplana septemlineata* was observed to predate upon *A. fulica* in Hawaii (Mead, 1963a). The occurrence of this predator appears to be restricted to the Hawaiian islands (Mead, 1963a). These worms are 4–6×0.2 cm in size and attack the snail in group. By intense sucking these can make deep holes and grooves at the site of attack. They show the "strongest preference" for the newly-hatched achatinas. This predator has its greatest effect in biological control by destroying the juveniles (Mead, 1963a). However, in spite of its "relatively small size" it will attack even the largest-sized *Achatina.* Mead (1963a) is of the view that amongst invertebrates, *Geoplana* is the only predator which can kill even the biggest-sized achatinas. He has observed a snail with 12.8 cm long shell and weighting 168 gm being predated upon by this worm, while the well known introduced molluscan predators, *Euglandina rosea* and *Gonaxis quadrilateralis* attack achatinas only up to 5 cm shell-size. These predatory snails also are attacked and killed by this worm.

The flat worms move into the snail's own slime trail, then glide over the shell/body and into the pneumostome and pallial cavity where kidney, lung and pericardium are lodged. These vital organs are damaged by the sucking and tearing action of the proboscis. If the snail withdraws its body in the shell, they attack the exposed parts of the mantle and foot and remove pieces of flesh from these. These worms have also been observed in the mass of eggs, feeding upon the hatching snails (Mead, 1979).

The worm is "strongly nocturnal", while during day time, not a single worm could be located; after dusk, from the same area the worms were seen emerging from leaf litter (Mead, 1979).

Geoplana vaga has been reported from California, and *G. mexicana* from California and Texas. Their role in the control of *A. fulica,* in Miami is not known (Mead, 1979). Ash (1967) recorded *G. forsterorum* in New Caledonia and in 1976 gave a detailed account of the role of this worm as predator of terrestrial gastropods and likely transmitter of the nematode, *Angiostrongylus cantonensis* to man.

Mead (1963a) has reported *G. septemlineata* to be the intermediate host of the nematode *A. cantonensis,* the causative organism of the frequently fatal eosinophilic meningoencephalitis in human beings. The tendency of this worm for resting on the lettuce leaves increases the chances of infection to human beings as is evident from a report of Mead (1963a) of the curled up worm on lettuce served in a restaurant. Thus, while *Geoplana* is an effective predator of *A. fulica,* there is a risk also in the use of this predator for the control of the snail.

In India so far there is no report of using *Geoplana* for the control of achatinas. Raut and Ghose (1979a) have reported the tubellarian *Bipalium indica* predating upon *A. fulica.* The predation was restricted to juveniles with 1.5–2.5 cm shell size. The worm crawls on a moving snail and supporting its body on the soil turns the snail upside down.

Then it reaches the upwardly directed shell aperture and secretes a toxic fluid over the exposed part of the snail to paralyse it. Later, the snail becomes inactive, the worm enters the shell and sucks the fluid for about 20-30 minutes. The snail does not become active again and later dies. The role of this worm, in spite of being an effective predator of the snail is "not significant" in the area where the worm was located because of its number being much less than that of the snail (Raut and Ghose, 1979a). So far, no information is available regarding *B. indica* being an intermediate host of the nematode *A. cantonensis* or carrier of any other human disease-causing organism.

(ii) *Arthropoda:* There are a number of predators of the giant African snail, belonging to different classes of this phylum as given below:

(a) Predatory millipede (Diplopoda)
(b) Centipedes (Chilopoda)
(c) Hermit crabs and the coconut crab (Crustacea) and
(d) Insects (Insecta)

(a) *Predatory Millipede*

The predatory millipede *Orthomorpha* sp. is endemic to South Andaman and is a natural enemy of the giant African snail. Generally millipedes are herbivorous. Certain millipedes do secrete poisonous chemicals like phenols (Eisner *et al.,* 1963a) and hydrocyanic acid (Eisner *et al.,* 1963b) from their stink glands but these secretions are reported to play a defensive role to keep off the predators of millipedes. So far *Orthomorpha* sp. (Fig. 7.2) is the only millipede known to be predatory. Secretions of the stink glands of this millipede have been tested by the sodium picrate and copper acetate – benzedine test and found to be hydrocyanic acid (Srivastava and Srivastava, 1967). It was observed that even big-sized snails fell victim to the attack of this millipede but these were never observed attacking the actively moving snails. When the snails were resting on the soil surface, the millipedes attacked from below many times. Since large number of millipedes attack the snail several times, secretion of their stink glands ultimately inactivates the snails to such an extent that they are not able to repulse the attacks and the millipedes start feeding on them by scraping their body surface (Fig. 7.3)

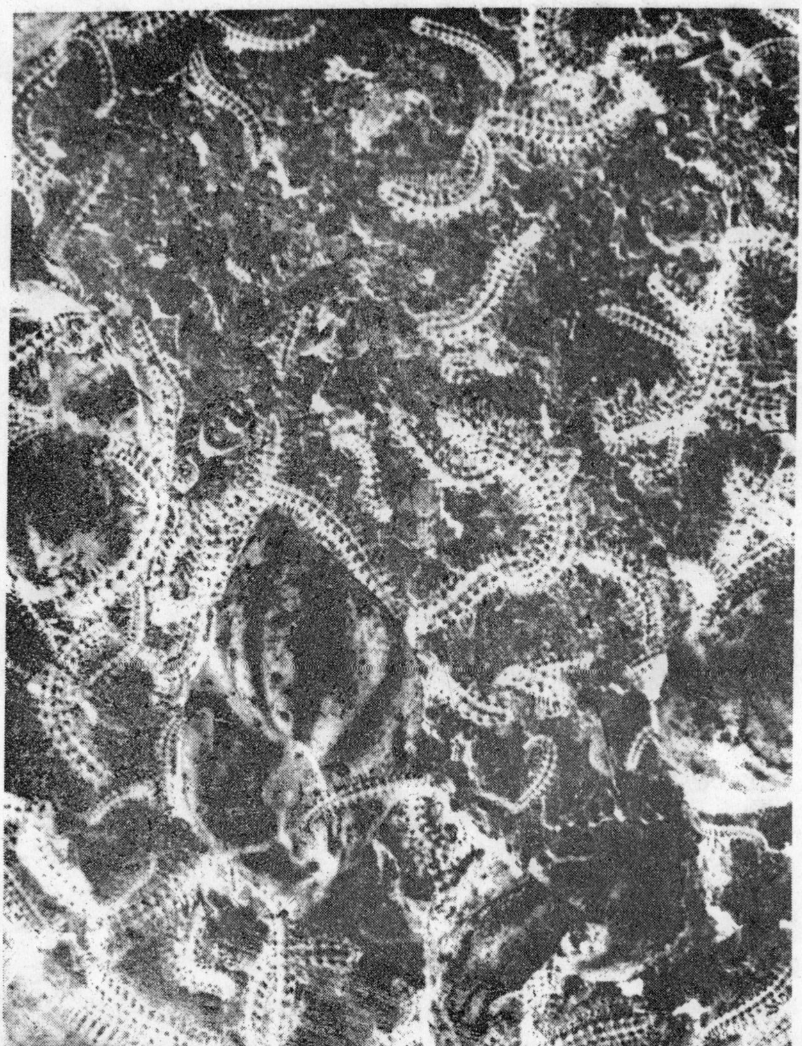

FIG. 7.2: *Orthomorpha* sp., the predatory millipede; openings of its stink glands visible

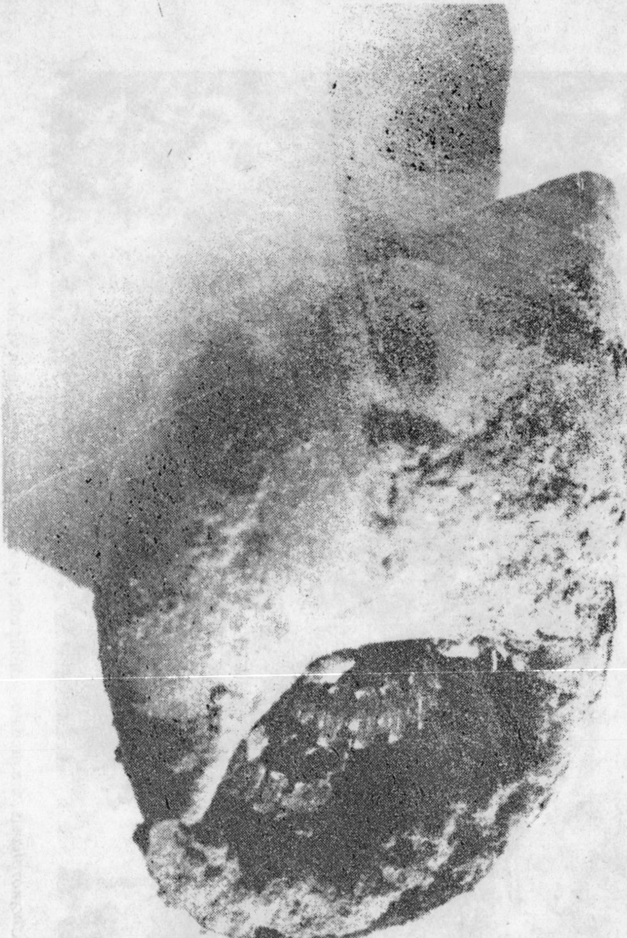

FIG. 7.3: Predatory millipede attacking a giant African snail

Orthomorpha sp. could be reared in the laboratory at the Indian Agricultural Research Institute, New Delhi. Since the breeding season of the millipede coincided with the breeding season of the snail, it could be easily used for the control of the snail.

An important aspect of the use of *Orthomorpha* sp. in the control of the snail is that the predator and the prey both belong to two different phyla, the predator belonging to phylum Arthropoda and the prey to Mollusca. This aspect is very significant because the chemical (metaldehyde) which is very effective in killing the snail does not kill its predator millipede. This millipede has been observed even nibbling the metaldehyde pellets without any adverse effect. Thus the predator and the chemical (metaldehyde) both can be used together in the field. (Pradhan and Srivastava, 1971).

Since breeding the predatory millipede in the laboratory is possible, laboratory-bred *Orthomorpha* sp. can be released in large numbers in small snail-infested pockets and in kitchen gardens. However, since large numbers are required for control, it may not be practically possible on field scale. However, it can be complementary to other control measures undertaken in the field.

(b) *Centipedes*

Vagvolgyi (1974) has listed the giant centipede as a predator of land snails. Mead (1979) reports that although the common giant centipede of the Western Pacific Islands has been observed in association with *A. fulica* but has not been observed to attack it; however, it has been observed feeding on some other small invertebrates (not snails). The predation of *Achatina* by the giant centipede needs further detailed research. On the Andamans also large number of large-sized centipedes have been seen. However, there are no reports of the centipede predating on the giant African snail. Even if one gets enough evidence of centipedes predating on the snails, it is not safe to rear these to increase the number as it injects poison from its poison glands through its poison claws, in the human body. The injection of poison by the centipede is very painful and sometimes even fatal. Under the circumstances, it is not feasible to make use of centipedes for the control of snails.

(c) *Hermit Crabs*

Hermit crabs of family Coenobitidae are amongst the most

conspicuous of the terrestrial fauna of tropical marine beaches.

In Micronesia Mead and Kondo (1949) and Mead (1950b,c) have observed *Caenobita perlatus* actually consuming the live snail by "pinching off small pieces of flesh with its chelae". Mead (1950b,c) has further observed the existence of an *Achatina* – free area about 3/4 mile (1.2 km) long and atleast 500 feet (152 m) wide along the east coast of Rota where the hermit crabs are "completely dominant". This *Achatina*—free area is bordered by "dense population of *A. fulica*". Somander (1951) describes unidentified hermit carb occupying *Achatina*-shells on the beach of eastern Ceylon, but he does not consider it as an act of predation. Davis (1954) says that "*Coenobita* was not a predator of *Achatina*". Mead (1961) feels that the conclusion of Davis is based largely on undependable behaviour of caged specimens. Van Driest (1968) has reported that he has seen hermit crabs in *Achatina* shells in Manokwari, West New Guinea.

Predation of *A. fulica* by hermit crabs (Fig. 7.4) was already known to scientists but the author was the first to use this crustacean for the control of *A. fulica*. In the words of Mead (1979) "Srivastava (1968 a) presents the best account ever made of the predatory potential of the hermit crab on the giant African snail".

On the main island of South Andaman 2000 numbers of *Coenobita* spp. (Fig. 7.5) were lodged in an enclosure in which different numbers of the giant African snail were released on four different dates (Table 7.1) and cent per cent mortality was observed in 1-4 days (Srivastava, 1968a). Thus after observing the effective predatory role of the hermit crab in the laboratory, the author moved to the field. In all, about 70,000 *Coenobita* spp. were collected from the sea beaches of South Andaman (Fig. 7.6) and released in several locations heavily infested with *A. fulica* during a period of about 2 months. This again "is the first instance" of employing this crustacean in biological control. This large scale release of hermit crabs in snail infested locations resulted in reduction of the snail population to such an extent that people who had given up vegetable cultivation because of snail menace started growing vegetables such as brinjal, okra and gourds. The hermit crabs were identified by the Zoological Survey of India, Calcutta as *Coenobita cavipes, C. perlata, C. clypeata* and *C. rugosa* (Srivastava, 1976).

Like the millipede *Orthomorpha* sp., hermit crabs also belong to phylum Arthropoda hence these also remain unaffected by the metaldehyde pellets which also can be used side by side (Pradhan and Srivastava, 1971).

TABLE 7.1: Small scale trial of predation of *A. fulica* by hermit crabs in a ventilated room of size 185×125×185 cm at Port Blair (Srivastava, 1968 a).

Date of starting the experiment	Date of observation	Total No. of snails Introduced	No. dead	No. inactive	No. active	%age mortality	Remarks
15-6-67	16-6-67	300	130	135	35	43.33	
	17-6-67	—	170	—	—	100%	
17-6-67	19-6-67	319	All the snails except 3 were floored down by the hermit crabs, mortality not noted down.				
	20-6-67	—	319	—	—	100%	Some shells were completely empty, some snails were partly eaten and dead, nine had entered snails' shells after killing and eating the snails contained therein.
20-6-67	22-6-67	94	94	—	—	100%	
23-6-67	25-6-67	254	120	134	-	47.24	
	27-6-67	—	134	—	—	100%	Six hermit crabs were observed to have entered snails' shells.

Hermit crabs can go up to 2.4 km away from the sea shore, where they are actually found, to find out suitable prey. Even in areas away from 2.4 km hermit crabs can be released after collecting from sea shore. In such cases most of the hermit crabs do remain there for about 4-6 months and carry on the predation of *A. fulica*.

Hermit crabs are found on sea shores but not on all the sea shores. It is difficult to explain why they are found on some shores and not on others. On South Andaman they are available in large numbers on the sea beaches of Burmah nullah and Cabynscove. However, the author could not find hermit crabs on the sea shores of Madras and Bombay. Wherever, these are not found, there are difficulties in using these as biological control agents. It would not be a practical proposition to use hermit crabs in inland areas like Bihar and Assam for the control of *Achatina*.

FIG. 7.4: The hermit crab *Coenobita* sp. — an effective predator of *A. fulica*

FIG. 7.5: Hermit crabs and *A. fulica* at Port Blair. Many of the snails have been already predated upon

FIG. 7.6: Hermit crabs being collected at a sea beach on South Andaman Island

The coconut crab or robber crab Birgus latro: This is a fairly big sized-crab. The young crab, like other hermit crabs, enters a gastropod shell for its protection and keeps on changing the shell as it grows but a stage comes when suitable shells suiting its size are not available, then it leaves it and starts living, without a protective shell. *B. latro* (Fig. 7.7) grows even up to 50 cm length and weighs up to 5 kg.

This crab is distributed over an extensive area in the Pacific and Indian Ocean islands. In India, it is reported to be prevalent on the Andaman and Nicobar group of Islands; even there, it is now available only on the South Sentinel island, other islands appear to have lost most of the coconut crab population (Davis, 1987). This crab is considered to be a predator of *A. fulica.* On the island of Auluptagel in the Palau group, Mead (1961) found over fifty broken shells of *Achatina* 3-4 feet (0.91-1.22 m) below the opening of the cave of a coconut crab indicating that these had been broken open by the crab. Kondo (1952) reports that on Agiguan, the coconut crab and rats together kill more snails than the introduced predatory snail *Gonaxis.* Davis's (1954) report of consumption of achatinas by the crab is based on direct and indirect evidences (Mead, 1961). Since predation of achatinas by the coconut crab is "only incidental", this crab plays the "smallest" role in the biological control of *Achatina* (Mead, 1961). In his revised edition of the book *The Giant African snail — A Problem in Economic Malacology* published as "Economic Malacology with particular reference to *Achatina fulica*" (1979), Mead has not referred to *Birgus latro* at all as a biological control agent of *A. fulica.*

(d) *Insects*

Ants: Schreurs (1963) has reported that red ants may kill *A. fulica* under unfavourable conditions i.e., in bright sunlight on dry soil but he himself states that these have no importance in the biological control of *A. fulica.* Dun (1967) states that the Fire Ant, *Solenopsis geminata rufa* partially immobilises snails by stinging them and then feeds on them. The victims are only those snails which are late to return to their hiding places. Such snails would have otherwise also died later in the day due to desiccation. Raut and Ghose (1984) have reported the ants of genus *Oecophyllus* carrying away the newly-hatched juvenile *Achatina.* However, there is no report of this ant attacking adult achatinas.

Beetles: Certain species of two coleopteran families, Carabidae and Lampyridae have been recorded as predators of *A. fulica.*

FIG. 7.7: The coconut crab, *Birgus latro,* another predator of *A. fulica* but also a pest of coconut

Carabidae: From this coleopteran family several species of genus *Tefflus* were introduced in Hawaiian islands. Of these *Tefflus zanzibaricus alluaudi* is a 4-5 cm long black-coloured beetle. Its adults and grubs, both are predacious. These were collected among fallen leaves etc. in the forest of Diani beach, Kenya. Releases of this beetle were made in the Hawaiian islands in 1952, 1953, 1957, 1959 and 1960 (Davis, 1960b, 1961c; Krauss, 1964). Some recoveries were made in 1958, 1960 and 1963 which created the impression that the beetle has probably established on the islands where these were released, but Davis and Chong (1969) report that no further recoveries have been made. Since mongoose in Hawaii is insectivorous, this may be the possible reason for no further recovery of this predatory beetle.

Other carabid beetles namely *T. purpureipennis wituensis, T. jamesoni, T. tenuicollis, Thermophilum hexastictum, Damaster blaptoides, D. blaptoides rugipennis* and *Scaphinotus* spp. also were released on Hawaiian islands but none of these could establish themselves.

Lampyridae: The Indian glow worm, *Lamprophorus tenebrosus* is endemic in India (including Andamans) and Sri Lanka. Mead (1979) considers it to be the "most spectacular" amongst those enlisted as predator of the giant African snail. Peterson (1957 a) states that during February and March 1955, six shipments of *L. tenebrosus* grubs were received from Ceylon (Sri Lanka) and released in different locations on Guam. The release areas were surveyed two months after the last release and there were some evidences to prove that some of the *L. tenebrosus* had survived. Empty shells of juvenile *Achatina* were fairly numerous at all the release points. In one area, a number of *L. tenebrosus* were observed feeding upon *Achatina.* Peterson (1957a) states that the release areas were surveyed several months after hibernation of the glow worms but there was no evidence of their establishment. Mead (1979) has referred to the release of *L. tenebrosus* in Guam, Indonesia, Mauritius and West New Guinea after getting shipments from Ceylon (Sri Lanka) in 1955, 1958 prior to 1955 and in 1959 respectively. However, there are no reports of establishment from any of these places. In Guam, no recoveries were made till 1972 (R. Muniappan, quoted by Mead, 1979). In Indonesia, these were probably fed upon by chickens (Dresner, 1958); in Mauritius, these did not establish (J. Monty, quoted by Mead, 1979) and in West New Guinea, all died (Schreurs, 1963; Van Driest, 1968). In West New Guinea, these died within a week of introduction. Mead (1979) is of the view that "rapid dramatic disappearance" of the specimens of even the largest releases seems to be because of the larvae being very sensitive to foreign environment even though the ecology of their endemic region and the site of release was very much the same. In addition, mammalian, avian and reptilian predators, the hermit crabs and other insect predators may also be responsible for the non-recovery of the released glow worms.

While the glow worms failed to establish wherever these were released, it is also a fact that *L. tenebrosus* continues to survive in Sri Lanka and India (including Andamans), the endemic areas of this predator including *Achatina*—infested areas. Thus they must be playing their role as a predator. What are the factors which are allowing the survival of the glow worm in its endemic areas and are these factors missing from the areas of its release — all these need thorough investigation.

No assessment of the extent to which this predator reduces the snail population in its endemic areas has been made. In spite the natural control being exercised by this predator on the snail population, the snail continues to be there causing economic injury to crops. If technique to rear this predator in the laboratory on mass scale could be developed and the snail infested areas in India (including Andaman) and Sri Lanka could be inundated with laboratory – bred glow worms, there are bright chances of the snail menace being reduced to a great extent.

Another lampyrid *Diaphanes* sp. was brought from Ceylon for release on Oahu, but all died while under observation in the laboratory (Thistle, 1957).

Yet another beetle predator *Selasia unicolor* belonging to family Drilidae was introduced in New Britain in 1963 after getting from Nigeria but it failed to establish (Mead, 1979).

Like the predatory millipede, *Orthomorpha* sp. and the hermit crab *Coenobita* spp., lampyrids and drilids also belong to phylum Arthropoda and their susceptibility to the molluscicide metaldehyde, is expected to be the same as that of the millipede and the hermit crab hence these also can be used along with the molluscicide metaldehyde.

(iii) *Predatory Snails*

Eleven species of predatory snails received from different countries were introduced into Hawaii between 1952 and 1961 for the control of

A. fulica. Out of these only four species namely *Euglandina rosea, Gonaxis quadrilateralis, G. kibweziensis* and *Gulella wahlbergii* could be recovered.

Euglandina rosea (Fig. 7.8) was introduced into Hawaii from Florida (U.S.A.) in 1955. Average size of the shell of mature snails was 5×2 cm. The eggs were white and their average size was 4.25×3-3.5 mm. Eggs were laid in clutches of 25–30 at the base of trees, in forest litter and on road side grass. Under quarantine laboratory conditions mature *E. rosae* fed on *Achatina* up to 4 cm shell size but preferred juveniles from 1.5-3.0 cm (Davis and Butler, 1964). This predatory snail was released in Hawaii between November, 1955 and March, 1956 at three different sites, the total number being 616. By 1957 more than 2483 *E. rosea* were recovered from Hauula and Makiki, Oahu and by 1958, over 12,000 were collected from these localities and distributed to other *Achatina*-infested locations. The population reached its peak in 1961 and it was considered to be the "most adaptable and abundant predator". Later its recovery went on declining and between 1955-71 it was reported to be "scarce". (Davis, 1972; Davis and Krauss, 1966). After the sharp population decline of *E. rosea* in Hawaii it has not yet recovered. However, on the basis of the remarkable spread of *E. rosea* in Hawaii, it seems unlikely that "once established this species would suddenly disappear" (Mead, 1979).

E. rosea was introduced in 20 more *Achatina*-infested sites of several countries. These have been reported to have "established" only at three places (Mead, 1979).

Gonaxis quadrilateralis: As early as in 1927 Pilsbry and Bequaert stated that perhaps the most important predaceous enemies of the African land molluscs were the snails belonging to family Streptaxidae to which *Gonaxis* belongs. This predatory snail was first introduced in Hawaii in 1957 from Kwale, Kenya. It is smaller than *E. rosea* in size. It is "without question, the most outstanding introduced enemy of *Achatina* in Hawaii" (Davis, 1961a). It has established in almost all the original 25 release points despite removal of over 12,000 from various release points and predation by rats and *Geoplana septemlineata* (Davis and Krauss, 1964). Mead (1979) also states that *G. quadrilateralis* has continued to manifest a gradual, strong and steady population increased and spread. *G. quadrilateralis* (Fig. 7.9) has a "definite preference" for the eggs of *A. fulica.* (Davis, 1960 a) but will feed readily on juvenile *Achatina* up to 35 mm in length, although it

FIG. 7.8: *Euglandina rosea*, an important snail predator of *A. fulica*

FIG. 7.9: *Gonaxis quadrilateralis*, another important snail predator of *A. fulica*

attacks larger specimen also. Davis and Krauss (1962) state that *G. quadrilateralis* "has definitely replaced *E. rosea* as the most promising predator of *A. fulica* in Hawaii and its establishment is unquestionately permanent".

Besides Hawaii this predator was also released in 14 more countries/islands between 1960 and 1974. Out of these 14 places it has been reported to have established in five places. In Mauritius it has been reported to have been recovered for the first time in 1970 after its release in 1962, while in Taiwan, after the release in 1967 the report is "often recovered in good condition". However, there is no statement of establishment in Mauritius and Taiwan. Reports for the rest of the places of release range from "establishment uncertain", "not given the results hoped for" to "no evidence of survival" (Mead, 1979).

When the Indian Agricultural Research Institute started work on *Achatina fulica* on the Andamans, it released *E. rosea* and *G. quadrilateralis* after getting from Hawaii, on two islands of the Andaman group. These predatory snails were kept in the Institute laboratory for about a month before airlifting and releasing on the islands. *E. rosea* laid eggs in the laboratory but no hatching was observed. *E. rosea* laid eggs at Rangat (Andaman) also in field cage, some of the eggs hatched also but could not survive to become adults. No egg laying was observed in *G. quadrilateralis.* A year after the releases no recoveries could be made. The releases were made twice in an effort to get established on the islands but the efforts could not succeed. There are unpublished reports of these predatory snails having been released on the mainland at Palghat (Kerala) and Baripada (Orissa) but there are no reports of their recoveries from these locations also. It is evident that the introduction of these predatory snails on the mainland of India and its islands was not successful and these could not establish in India (including Andamans). Besides other reasons it is possible that in spite of the ecological conditions of *Achatina*-infested locations of the Indian region and those of Hawaii from where these predatory snails were procured, being similar, these also may be susceptible to the foreign environment in the same way as the Indian glow worm collected from Sri Lanka and released in different places including the Hawaiian islands failed to establish there.

Gonaxis kibweziensis: It is similar in appearance to *G. quadrilateralis* but is about half the size of the former. It attains the "maximum" length of 2.2 cm (Davis and Butler, 1964). The shell is whitish and is conspicuously grooved (Williams, 1951). Like *G. quadrilateralis* this species also feeds on the eggs of *Achatina* and has been found "almost constantly feeding on infants and juveniles of *Achatina* up to 3.5 cm shell size. Davis and Butler (1964) observed 12 adults from a clutch of 200 to 400 eggs.

It was first introduced in Hawaii in 1952 in Kaneohe (Oahu) and later in other localities on Oahu and Maui in 1954, 1955 and 1957 (Davis and Butler, 1964). Davis (1958) states that this predator's "performance to date under Hawaiian conditions has been disappointing".

This predatory snail was introduced in a number of other countries. It established in all the places of introduction except in West New Guinea where all were reported to have died. Although this species has become established in Hawaii but its development has continued to be slow. Its cannibalistic habit may be reason for this (Davis and Butler, 1964). Mead (1979) ranks it as a "poor third" as a predator of *A. fulica* when compared with *G. quadrilateralis* and *E. rosea.*

Gulella: Gulella spp., *Gulella wahlbergii* and *G. bicolor* have been reported to be predating on *Achatina* in a number of *Achatina*-infested areas. These have been collected and released in several *A. fulica*—infested areas.

Gulella spp. were collected from Durban and Natal (South Africa) and Yaugambi (Congo) and released on Hawaiian islands (Krauss, 1964). However, there are no reports of their successful establishment at their release points.

G. wahlbergii were collected among fallen leaves and humus from Durban (South Africa) in 1956-57 and released on Oahu and Maui. Recoveries were made in 1960 and 1964. Davis and Krauss (1964) are of the view that because of its very small size, it is "unlikely to be of any importance" in the control of *A. fulica.*

G. bicolor also is a small-sized predatory snail. It was collected from Manila (Philippines), Susupe, Saipan, (Mariana islands) and Kuala Lumpur (Malaysia) in 1957-58. These were released on Oahu in 1957-58. Mead (1961) states that *G. bicolor* has "proved to be of negative value" so far as predating on *Achatina* is concerned.

The only predatory snail reported from India is *Gulella (Indoennea) bicolor* reported by Annandale and Prasad (1920) as *Ennea bicolor* predating on *Opeas* sp. — a pestiferous land snail. This predator has been reported from three small islands, Chattam, Ross and Viper of South Andamans. Godwin-Austen (1908) have stated that Mr. Collett observed the association of *Ennea bicolor* and *Opeas* sp. in Ceylon. This predatory snail was reported by the author in 1967 for the first time from Delhi (Srivastava, 1968b). This was identified by the Zoological Survey of India as *Gulella (Indoennea) bicolor*. They also remarked that most of the species were described by earlier workers as *Ennea*. Currently, *Ennea bicolor* is treated as *Gulella (Indoennea) bicolor* Hutton (N.V. Subba Rao : personal communication, 14-9-67). *Gulella (Indoennea) bicolor* was collected from Delhi mainly from the Indian Agricultural Research Institute Campus and released in an *Achatina*—infested pocket of Calicut (South Andaman).

The ratio of *Gulella* and *Opeas* in Delhi has been found to be 1 : 9. In the laboratory *Opeas gracile* and juveniles of *Achatina* (shell-size 5–15 mm) were offered to it in different combinations with replications and checks. When only juvenile *Achatina* were kept with *Gulella*, some were eaten up; when only *Opeas* was offered they were not eaten atleast for 5 to 7 days. When *Opeas* and *Achatina* juveniles both were kept together in equal numbers to test the predation of *Gulella*, the predation of *Achatina* and *Opeas* was observed to be 23 to 60 per cent and 0-6.7 per cent respectively. Thus *Gulella* cannot be considered as an effective predator, either of *Achatina* juveniles or of *Opeas* and that when *Achatina* juveniles and *Opeas* both are available, it shows distinct preference for *Achatina* (Srivastava *et al.*, 1975). Raut and Ghose (1984) have observed *Gulella* predating on *Opeas* but not on *Achatina* in the field. The author has observed this predator predating on *Achatina* in the laboratory. Further work is needed to establish distinct predation preference of *Gulella (Indoennea) bicolor* vis-a-vis pestiferous land snails.

However, because of its small size it does not appear possible that it would ever prove to be an effective predator of *A. fulica*, unless the *Achatina*-infested pockets are inundated with laboratory-reared *Gulella*.

The pest *A. fulica* and the predatory snails belong to the same phylum Mollusca. Thus while using snail predators to contain the population of the pest snail, no chemical can be used for snail control as the chemicals which will kill the pest snails, will kill the predatory snails also. Since hitherto there is no record of *A. fulica* having been controlled fully by any predator, we have to continue to use chemicals also for its control and wherever chemical is used, predatory snails cannot be used.

(B) *Vertebrate Predators*

(a) *Birds*: Amongst birds, only nocturnal birds like owls can be effective predators. Mead (1979) has remarked that most Biologists consider importation of birds as "precarious and impractical" except the release of the barn owls (*Tyto alba*) and the cattle egrets (*Bubulcus ibis*) on Hawaiian islands (Davis, 1960b; Davis, 1961a; Davis and Krauss, 1962).

Raut and Ghose (1979b) have reported the Tree pie (*Dendrocitta vagabunda*) and the crow pheasant (*Centropus siensis*) as natural enemies of *A. fulica*. However, they have also reported that these birds predate upon the snails only during day time when the snails are seen moving about after a shower or when the sky is overcast. Since predation of snails by these birds is restricted only to a few days in a year, these birds are not very effective in reducing the snail population (Raut and Ghose, 1979 b).

Common ducks were observed by the author on the Andaman islands to feed on the eggs and juvenile *Achatina* up to 3 cm shell-size; however, they prefer the juveniles up to 1.5 cm shell-size as their shell is very soft. Certain farmers and others keep the ducks to keep a check on the eggs and juveniles (Srivastava *et al.,* 1970). If migration from the nearby areas could be checked, the snail population is bound to decline. Because of the predation of eggs and juveniles these would not be allowed to become adults and those already present would ultimately become unproductive and die in course of time. Peterson (1957b) also has stated that "ducks feed readily on living *Achatina* especially juveniles". He has also reported that certain farmers crushed the snails and fed to their ducks; crushed and boiled snails were "readily acceptable to chickens".

(b) *Mammals*: The musk shrew (*Suncus marinus*) became established on Guam in two years, but observations both in the field as well as in the laboratory did not give nay indication of its being a potent biological control agent of *A. fulica* (Peterson, 1957b). However, Nakata (1960) has observed the musk shrew feeding on "living *Achatina*".

The bandicoot rat (*Bandicota indica*) has been reported by Raut and Ghose (1979b) as an "effective predator" of the giant African snail as it can locate the active as well as aestivating snails. These authors suggest that the bandicoot rats can be effectively used in snail control but they also caution that its population density should not cross a level at which the predator itself becomes a major pest. Landry (1970) reports that besides rats and mice certain species of ground squirrels also are malacophagous.

South (1926) has listed wild pig as a predator of *A. fulica*. Mead (1961) states that omnivorousness of domestic pigs "suggests correctly" that they would feed upon snails near habitations. The author failed to feed the giant African snail to the pigs at Delhi; however, during one of his visits to Andaman he observed a resident in Calicut (Andaman) collecting snails for feeding to his pigs (Srivastava *et al.*, 1971). In Andaman, the local breed of pig *sus andamanensis* was found to be a voracious feeder of *Achatina* as it could eat, on an average, 260 snails (small to medium-sized) i.e., about 3.80 kg every day (Fig. 7.10).

Rats including bandicoots, squirrels and pigs, all these are crop pests also. How far can these be really effective in reducing snail population without damaging the crop is difficult to assess and in nature it may not be possible to depend upon these mammalian predators cum pests.

The mongoose, *Herpestes mungo* is reported by Philbrick (1949) and Rees (1950) to be feeding on the giant African snail in Ceylon (Sri Lanka). Mead (1961) has also stated that he got sufficient circumstantial evidence in Ceylon in this regard. Mongoose is a well known predatory mammal and is reported to be usually predating on rats. This predator is quite common in India also and if their number could be increased in the snail-infested locations by collecting from non-snail-infested areas and releasing in snail-infested pockets, it might prove helpful in reducing the snail population.

FIG. 7.10: *Sus andamanesis,* the Andaman pig feeding on snails

Parasites

The main parasites of giant African snail are parasitic Protozoa, Cestodes, Trematodes, Nematodes and flies of family Sciomyzidae. Protozoans have been reported from terrestrial gastropods; however, Richards (1968) lays stress on conducting intensive research into the use of protozoans as biological control agents. Mead (1979) states that the potential of Protozoa as biological control agents for controlling *A. fulica* is very uncertain. Raut (1980) has reported a ciliate *Trichodina* sp. from the spermathecal fluid of *A. fulica.* The infection was only 8 per cent in young snails, 17 per cent in the juveniles and 44 per cent in the mature ones. However the role of this ciliate in reducing the snail population is yet uncertain, as it is yet to be determined whether the ciliate is parasitic or symbiotic.

(a) *Cestodes and Trematodes:* In addition to snail, cestodes have a vertebrate host also so their use as biological control agents of *Achatina* and other land snails does not appear to be a practical proposition. Similarly snails also act as intermediate hosts of trematodes and are not known to cause mortality of *A. fulica.*

(b) *Nematodes:* Just as we are not yet in a position to use cestodes and trematodes for the control of *Achatina* and other land snails, in the same way, we are not yet in a position to use parasitic nematodes as effective biological control agents for controlling the population of *A. fulica.*

Seventy seven species of nematodes have been found associated with land snails; out of these, 55 have some vertebrate host also. So far only two species namely *Anafilaroides rostratus* and *Angiostrongylus cantonensis* have been identified from *A. fulica* (Mead, 1979). Raut and Ghose (1984) have recorded a rhabditoid nematode from the gut of about 35 per cent *A. fulica* and another land snail *Macrochlamys indica.* Thus we are not yet in a position to exploit nematodes for the control of *A. fulica:* Thorough study of the possibility of using nematodes for the control of terrestrial gastropods is necessary (Mead, 1979).

(c) *Flies:* Flies of family Sciomyzidae are reported to be parasitic on pulmonates. *Sciomyza dorasta* reported from Denmark might prove useful in the control of *A. fulica*. However, possibility of the introduced predatory snails *E. rosea, G. quadrilateralis* and *G. kibweziensis* being affected by this parasitic fly is a cause of concern. However, Davis (1961 b) reported no danger of *S. dorsata* to the predatory snails.

(d) *Pathogens*: Annandale (1919) of India was the first Scientist to report that *A. fulica* was subject to some kind of "fatal epidemic". Taking a clue from this report, South (1926) reported this "fatal epidemic" as a controlling factor in the biology of *A. fulica*. Later, D.B., Langford (1947, quoted by Mead, 1961) suspected disease in *A. fulica* on the Trust Territory of the Pacific Islands; Muma (1954) reported the possibility of a bacterial disease in the tree snail *Drymaeus dormanii* in Florida, C. Vago (quoted by Mead, 1961) reported from France in 1955, the death of about 50,000 snails in "Snail Parks" where they are raised in close confinement for human consumption.

In Ceylon in some areas, a serious build up of *Achatina* population in 1951 and a sharp decline in 1952 "without undertaking any control measures" was reported. When some specimen from the field were examined in the laboratory, quite a number of specimen were found to have leukodermic lesions on the tentacles and occasionally on the face and neck also. The lesions may become elevated into tuberculations (Mead, 1961). Van Zwaluwenburg (1955) observed leukodermic lesions on *A. fulica* in Singapore, Hong Kong and Hawaii while Mead (1958a,b) reported diseased snails from Bangkok and Thailand in 1957; in contrast to the Hawaiian specimens, leukodermia was much less pronounced but the tuberculations were severe.

Mead (1969) reported *Aeromonas liquefaciens* to be responsible for the leukodermic disease in *A. fulica*. In 1970, Dean *et al.*, (1970) isolated the bactermium *Aeromonas liquefaciens* from the diseased *A. fulica* in Hawaii. Sturtevant (1969) is of the view that taxonomically *A. liquefaciens* "probably" should be referred to as *A. hydrophila*.

In India, Srivastava (1966) observed "Leucodermia-like disease" (Fig. 7.11) in April, 1965 in the laboratory culture of the Indian Agricultural Research Institute, New Delhi. The symptoms were the same as reported by Van Zwaluwenburg (1955) and Mead (1958 a, b).

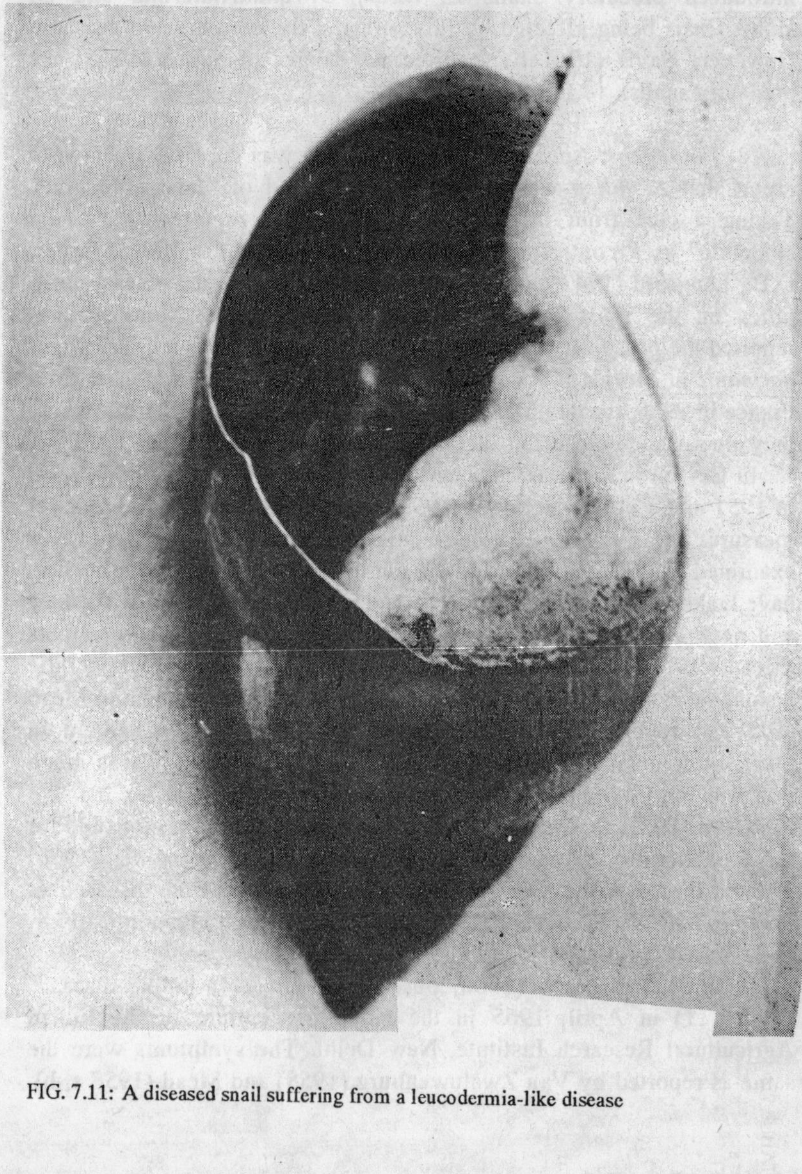

FIG. 7.11: A diseased snail suffering from a leucodermia-like disease

TABLE 7.2: Field-cage trials with the diseased snails' extract spray at Port Blair (Srivastava and Srivastava, 1968)

Sl. No.	Number of diseased snails crushed	Date of spraying	Date of observation	Total number of healthy snails introduced	Number of dead snails	Number of living snails	% mortality	Remarks
1	6	9.6.1957	13.6.1957	1650	1350	300	81.8	In control (untreated) snails were kept under similar environmental conditions but no mortality was observed during the course of the trial.
			16.6.1967	0	220	80	73.5	
			18.6.1967	0	80	0	100	
2	2	19.6.1967	23.6.1967	130	127	53	70.5	
			25.6.1967	0	53	0	100	

To begin with, the snails became sluggish, suspended feeding and ultimately died. A study of the dead/dying specimens revealed the existence of tuberculations on the foot, and a number of pigmentless scars were also observed on the diseased snails. When the disease was located, the rearing room was thickly populated and the humidity and temperature both were quite high. Mead (1961) has stated that the high moisture prevailing during the snail's activity period was favourable for the development of the disease also. Raut and Ghose (1977) reported the outbreak of Leucodermia-like disease in *A. fulica* from West Bengal. In the field population, they recorded 47 per cent infection and 50 per cent in experimental cages. They have reported 25-34°C and 90-96 per cent temperature and humidi, respectively to be responsible for the outbreak of the disease. Raut (1983 b) observed the disease syndrome in some snail-infested locations in West Bengal, Assam, Nagaland, Manipur, Bihar, Orissa, Uttar Pradesh, Tamil Nadu and Kerala.

Srivastava and Srivastava (1968) sprayed the aqueous extract of diseased snail on the field population in three localities of South Andaman. The disease was successfully spread in two localities; in the third locality spraying was followed by heavy showers hence, the sprayed material was washed away by rain.

In some places, the cause of sudden population decline of *A. fulica* is considered to be the disease. Now since the causative organism has been identified, if it could be cultured in the laboratory on large scale with the help of bacteriologists, and technique standardised for spreading the disease in the field, it could go a long way in tackling the giant African snail problem, in all the countries of the world where this snail is a problem.

Chemical Control

It is true that biological control is the ideal method of control of snails. However, since researches in this direction have not yet reached a stage when we may depend solely on biological control, the cultivator cannot wait hence we have to take the help of chemicals.

While selecting the chemical the following points have to be taken into consideration:

1. The chemical should be able to affect heavy mortality of the snail.

2. The chemical should have no adverse effect on soil.

3. The chemical should have negligible or no harmful effect on the fauna (including soil fauna) and flora.

4. Should not be toxic to warm-blooded animals especially to poultry, cattle, sheep, goats and to human beings.

5. Since snails are normally active during the rainy season, the chemical used should be such that it may not get dissolved and washed away by rain. Hence in the field, effect of dusts and sprays is short lived because once it rains dusted and sprayed material gets washed away. So, as far as possible, the chemical should be used in the form of stable bait which may continue to remain effective for several days and may not get washed away by rain.

However, it is difficult to find a chemical with all the qualities, but effort should be made to use a chemical with as many qualities as possible.

In the snail-infested countries, including India, a number of chemicals and their different formulations have been used. The chemicals generally used have been the arsenicals, copper compounds, common salt (Sodium chloride), chlorinated hydrocarbons. organophosphates, carbamates and metaldehyde.

Arsenicals
1

*Calcium arsenate:** Several workers have tested this chemical against a number of land snail pests and found it effective also. Some have

1. * Now not permitted to be registered in India.

tested this chemical against the giant African snail also. Lange (1950) used pellets containing metaldehyde and calcium arsenate on Saipan. Peterson (1957b) used a commercial bait in the form of pellet prepared by mixing bran, 1.6 per cent metaldehyde and 5 per cent calcium arsenate which "proved highly effective and relatively safe". However, the pellets disintegrated soon during the rainy season. Thus pellets were kept on a pan under a shelter of boards. He states that the snails got paralysed in 10-15 minutes and died in half to one hour. If the snails thus killed, were eaten by other snails, they also died.

Peterson (1957b) also tried poisonous white-wash on the fence wall in Guam with a preparation containing 40 per cent calcium arsenate, 55 per cent slaked lime and 5 per cent cement mixed with enough water to make a thin paste. This method, however, proved rather ineffective as the surface of the ground over much of Guam consists mainly of corals hence snails get abundant calcium and did not get attracted to the poisoned white wash. However, Rees (1951) recommended the spraying of lime covered walls with 1 per cent solution of this chemical. Beeley (1938) recommended a bait of calcium arsenate, slaked lime and cement in the ratio of 1 : 6 : 2 (V/V) or 1 : 4 : 2 (W/W) and enough water to form an ordinary concrete mix. When dried, the slab was broken into small pieces and scattered in the snail infested locations. Fullaway (1949) recommended and used Beeley's bait ingredients for several years in Hawaii in the ratio of 8 : 11 : 1 (W/W) to which water was added to make a thin paste. This paste was painted on rocks, tree trunks and such other objects with which the snail was likely to come in contact. However, bran, 1.6 per cent metaldehyde and 5 per cent calcium arsenate bait in the form of pellet "proved highly effective" against *A. fulica* (Peterson, 1957 b).

In India, Mandal and Ghose (1970) have reported cent per cent mortality of *A. fulica*, within 65-70 hours of dusting with calcium arsenate. Signs of lesions were detectable within 5-6 hours after dusting on the exposed portions of the body which came in contact with the chemical. Bhardwaj (1972) prepared a bait of calcium arsenate and bran in the ratio of 1 : 4 (W/W), tested against *A. fulica* under field-caged condition and got 92 per cent mortality within 120 hours after the treatment (Table 7.3).

Besides calcium arsenate, other arsenicals used against land snails including *A. fulica* are calcium arsenate, lead arsenate, sodium arsenate, sodium arsenate and white arsenic. In some cases these have proved effective, in others ineffective. All arsenicals are highly toxic stomach poisons and are toxic to warm-blooded animals also. Thus arsenicals have slowly phased out as pesticides.

TABLE 7.3: Field-cage trials with different formulations of pesticides against *A. fulica* at IARI, New Delhi (Bhardwaj, 1972)

Pesticide	Concentration (%)	Percentage mortality 120 hr after treatment
Endrin* (dusting)	1.5	28.0
	1.0	20.0
Copper sulphate (dusting)	40.0	95.8
	35.0	83.3
	30.0	66.7
	25.0	58.3
	20.0	41.7
	15.0	29.2
Calcium arsenate** (bait)	20.0	92.0
	15.0	53.2
Metaldehyde (bait)	5.0	96.7
	3.0	50.0

* Now its use in India is prohibited.

** Now not permitted to be registered in India.

Copper Compounds

Copper sulphate: Amongst the copper compounds, copper sulphate has been generally used to control the giant African snail. Since 1910, Green (1910), Hutson (1920), South (1926), Corbett (1933) and Fernando (1952) have recommended its use as dust, spray and bait. Pangga (1949) used 1 to 10 per cent solutions of copper sulphate against this snail. He states that although these solutions killed the young snails "readily" but not many adult snails. Peterson (1957b) reports that the areas not enclosed by wall etc. could get temporary protection from *Achatina* by "a line of wood ash about 6 inches (15 cm) wide and 1-2 inches (2.5–5 cm) deep" over which was sprayed 10 per cent solution of $CuSO_4$. This acted as a barrier. Saxena and Dubey

(1970) used $CuSO_4$ dust @ 5.5 kg/ha against *A. fulica* and recorded 93 per cent mortality (Table 7.4), while Bhardwaj (1972) reported 95.8 per cent mortality of this snail within 120 hours after dusting with a mixture of copper sulphate and lime in the ratio of 40 : 60 in his small scale 'field cage' experiments (Table 7.3). Snail-infested areas are prone to heavy rains; the dusted chemical is washed away and dusting has to be done over and over again which is obviously uneconomical. Moreover, continued use of copper sulphate is likely to increase acidity of soil, which may be harmful to crops which are susceptible to acidity. In addition, plants which have been injured due to the feeding by *A. fulica* absorb $CuSO_4$ rapidly at the site of injury and die. Copper sulphate mixed with lime also has been recommended for dusting over the snail in the ratio of 1 : 4 or 1 : 5. This reduces the chances of increase in the acidity of soils. Dusting of $CuSO_4$+ lime is reported to have been done by some scientists of the Directorate of Plant Protection, Government of India on the Andamans in the late sixties. Copper sulphate and lime, both being hygroscopic, the nozzle of duster got clogged frequently. Dusting of copper sulphate over snails in Egypt to kill snails responsible for the disease Bilharziasis among human beings resulted in the heavy mortality of Sardine fishes which form 60 per cent of fish fauna of the mediterranean region.

Common Salt

Sodium chloride: Sodium chloride has been used by scientists in India as well as in foreign countries for the control of snails and slugs. It has been used as a spray as well as in crystal form. Like copper sulphate and lime, sodium chloride crystals also are hygroscopic therefore when the crystals fall on the muscular part of the snail, it dies after some time, due to dehydration.

Latif (1933) used sodium chloride to protect potted orchids from *A. fulica*. Hall (1932) advocated the use of aqueous solution of NaCl instead of crystals, but Pangga (1949) reported the mortality of young *A. fulica* with salt solution. Peterson (1957b) found salt water spray either prepared from sodium chloride or as sea water to be effective in killing snails in Guam, but it caused burning of foliage. He considered such sprays to be of some practical utility only in sea beach areas and along side of roads where damage to vegetation or soil was "of no consequence".

In India, Singh and Birat (1969) have reported the dusting of common salt on the crawling *A. fulica*. Saxena and Dubey (1970) dusted common salt @ 200 kg/ha and got 98 per cent mortality of *A. fulica* in the field six hours after the treatment (Table 7.4). While Peterson (1957b) considers even spraying of salt water to be only of "some practical utility", in sea beach areas and on road sides and it caused burning of leaves, Saxena and Dubey (1970) used such a heavy dose as 200 kg/ha, of common salt and reported 98 per cent mortality of *A. fulica,* six hours after the treatment without considering the fate of the soil. It is surprising that while Peterson (1957b) has reported burning of leaves even with salt water spray such a heavy dose as 200 kg/ha did not cause any burning of the crop. What is the purpose of using such a heavy dose of common salt, which, while killing the snails ruins the soil ? These authors should have taken the advice of Central Soil Salinity Research Institute, Karnal (Haryana) to know the fate of soil with such a treatment. Raut and Ghose (1984) also have reported the use of common salt in the Eastern region of the country to kill *A. fulica.*

TABLE 7.4: Field trails with chemicals against *A. fulica* (After Saxena and Dubey, 1970)

Name of the chemical with formulation	Rate of application per hectare	Mortality (percentage)	Effect on grass
Metaldehyde with herbaceous base (as bait)	200 kg	98	Nil
Metaldehyde with some other base (as bait)	200 kg	89	Nil
Sevin (as 10% dust)	100 kg	98	Nil
	40 kg	78	Nil
Copper sulphate (as dust)	11.0 kg	98	Slight effect on leaves
	5.5 kg	93	Nil
Common salt (as dust)	430 kg	100	Completely burnt
	200 kg	98	Nil
Mixture of 2,4-D and 2,3, 5-T (as spray) (2 : 1)	50 l.	91	Completely burnt
	27.5 l.	78	Only leaves burnt

Chlorinated Hydrocarbons

Benzene Haxachloride (BHC) : BHC has been used very little for the control of snails. As regards the results there are conflicting reports.

Pereira and Goncalves (1949) have reported BHC to be "inferior to metaldehyde" for the control of snails and slugs. Rao *et al.,* (1953) dusted 5 per cent BHC in a drained paddy field @ 62 pounds/ha and which "killed most of the snails in the field". He found BHC to be superior to 2 per cent metaldehyde bait. The snails in the paddy field, as expected, could not have been land snails. Pappas and Carman (1955), working on *Helix aspersa* found this pesticide to be "slightly less than moderately effective" in killing *H. aspersa.* Some scientists have used the gamma isomer of this chemical for controlling snails and slugs. Fromming and Riemschneider (1952) found it to be ineffective against slugs. Pangga (1949) found it to be "ineffective" against *A. fulica* while Tubb (1952, quoted by Mead, 1961) found it to be "effective" against the giant African snail in North Borneo.

Before the Indian Agricultural Research Institute, New Delhi started working on *A. fulica* on the Andamans in 1964, the Department of Agriculture, had been using BHC and DDT for the control of this pest without any success. The author has himself seen this snail crawling over BHC spread on the pathways of the snail, moving forward and getting busy in its biological activities. Crowell (1967) has reported little or no toxicity by the chlorinated hydrocarbons and organophosphates tested against snails and slugs.

Recently Jagtap *et al.,* (1989) reported 5 per cent BHC bait to give 81.11 per cent mortality of the land snail *Cryptozona belangiri.* There does not appear to be any earlier published report of the use of BHC as bait against snails and slugs.

Since snails are abundant during the rainy season, dusting or spraying are normally not considered to be useful on field scale as these get washed away in no time. Bait appears to be the only formulation useful in the field, but even bait has to be shower proof and a little heavier so that these may not get dissolved or washed away when it rains. Addition of cement/lime would make the baits heavier.

It is possible that BHC may be effective against certain species of snails under certain circumstances. However, neither Mead (1961, 1979) nor Raut and Ghose (1984), who have done exhaustive studies on land snails, have recommended BHC for the control of *A. fulica* and other land snails. The author and his co-workers, after working on land snails and slugs for about 16 years, have recommended the use of 5 per cent metaldehyde bait and not BHC. Before recommending BHC for the control of snail, work on various aspects has to be done, which, at present, is not being done anywhere, atleast in India.

DDT: Very few experiments have been performed on the possible use of DDT as a molluscicide. Buckhurst (1947) had initially got "slightly encouraging results" with 0.1 per cent, emulsion, 5 per cent dust and as bait. Slugs did not even touch the bait. When he tested 0.2 per cent emulsion he got "an excellent control of slugs" without any phytotoxicity to the cabbage plants. Thomas (1948) tested DDT-bran bait against slugs and observed that although the slugs were attracted to the bait (because of the bran) they were not killed by the bait.

Fromming (1949, 1950), Fromming and Riemschneider (1952) and Fromming and Plate (1952) did not find any molluscicidal effect of DDT. Like-wise Hely (1946), Pappas and Carman (1955) Pereira and Goncalves (1949) and Karlin and Naegele (1958) could not recommend it against snail and slugs.

Dieldrin: Lange (1952), (quoted by Mead, 1961) found DDT to be of relatively little value as molluscicide even in combination with metaldehyde, when used against *H. aspersa.* While Pappas and Carman (1955) agree with Lange, Karlin and Naegele (1958) found it to be ineffective against slugs.

Endrin: Manna and Ghose (1972) found 2 per cent endrin bait to give cent per cent mortality of *A. fulica.* Bhardwaj (1972) tested 1.0 and 1.5 per cent dust against *A. fulica* in the field cage at Delhi during July-August when the relative humidity varied from 80 to 97 per cent. The highest mortality obtained was 28.0 per cent from 1.5 per cent dust, 120 hours after the treatment (Table 7.3). Verma (1965) has reported endrin to be less effective than metaldehyde. Thus endrin also, like BHC, DDT and dieldrin is not a suitable molluscicide. Therefore, chlorinated hydrocarbons cannot be recommended as molluscicides for the control of snails including *A. fulica.*

Organophosphates

Organophosphates have not been used much as these have not been found to be very effective against snails.

Manna (1976) has tested diazinon and fenitrothion. He used these two chemicals in the form of bait. Baits were prepared with bran, molasses and the chemical.

Two per cent and one per cent diazinon baits gave 70 and 60 per cent kill of *A. fulica* respectively. Two per cent fenitrothion bait gave 50 per cent kill.

Jagtap *et al.,* (1989) as a result of three year trials got 25.55 per cent mortality of *Cryptozona belangiri* with phorate used @ 1.5 kg ai/ha in the form of bait.

Carbamates

Amongst carbamates Temic (*aldicarb*), Sevin (*carbaryl*) and Mesurol (*Methiocarb*) have been found useful for the control of land snails and slugs. Judge (1969) obtained spectacular results with Temic in the control of slugs but he also states that it is "dangerously toxic" by all routes of exposure. Wouters (1970) and Symonds (1975) have reported mortality of snails and slugs by Mesurol but stated that it kills earthworms also. Wouters (1970) found mesurol to have a higher molluscicidal effect on slugs than metaldehyde but Gorokhov (1971) reported "opposite results". Mesurol was tested in Mauritius in 1971 against *A. fulica* and is still being used along with metaldehyde (Mead, 1979). Crowell (1967) has reported the use of carbamates especially mesurol in, the form of bait to be useful for the control of land snails. Bhat and Shamanna (1973) reported 100 per cent mortality of land snails in South India with 5 per cent sevin as bait. Saxena and Dubey (1970) got 98 and 84 per cent kill of *A. fulica* with 10 per cent sevin dust used @ 100 and 60 kg/ha respectively. Jagtap *et al.*, (1989) have reported, as an average of three years trails, 51.18 per cent kill of *C. belangiri* by 5 per cent sevin bait and 28.88 per cent kill by carbofuran when used @ 1.5 kg ai/ha as bait.

Olson (1973) screened a large number of pesticides including 46 carbamates against *A. fulica.* the carbamate compounds Shell's SD-17250 and the Union Carbide's μC-30045 produced consistently high mortality of *A. fulica.* After the report of Hunter and Symonds (1970) that a combination of metaldehyde and the carbamate methiocarb produced high slug kills, it is now felt that pellets of either methiocarb or metaldehyde will give about equally high kill of terrestrial gastropods, but "methiocarb is more toxic to the environment" (Gould and Webley, 1972).

Metaldehyde

Metaldehyde was used for the first time as a molluscicide in South Africa in 1934 (Gimingham, 1940) whereas in England it was used as a molluscicide in 1936 (Jary, 1939).

Metaldehyde is an inflammable polymerized form of acetaldehyde having the chemical formula $(CH_3CHO)_4$. Its solubility in water is very low. At 30°C the solubility is 0.026 per cent but at 17°C it is only 0.020 per cent. Even after standing for more than six months, no tendency to depolymerize in neutral or alkaline aqueous solutions, has been detected (Cragg and Vincent, 1952).

As regards the action of metaldehyde on molluscs, many feel that it is a specific attractant. Corbett and Pagden (1941) observed that *A. fulica* will crawl towards bait containing metaldehyde and then change the course just before reaching it. Weber (1954) reported that dry metaldehyde and metaldehyde "in solution" failed to attract *A. fulica.*

Jary and Austin (1937) were the first to conclude that metaldehyde is both contact as well as stomach poison. Cameron (1939) and Lange and MacLeod (1941) agreed with Jary and Austin (1937) but Thomas (1948) was of the view that metaldehyde was only a stomach poison, as far as snails are concerned. However, he referred to three types of actions on slugs : (i) an instant effect due to which there is excessive secretion of mucus. If the slug is washed with water and kept away from desiccating air, it would recover, (ii) an anaesthetic effect causing complete immobility and (iii) an irreversible toxic effect and producing a characteristic transparency in the gut. Cragg and Vincent (1952) demonstrated that metaldehyde is both contact as well as stomach poison, and has a 'progressively greater toxic action' in the haemocoele, in the crop and on the surface of the body. They did not observe any transparency of the gut wall nor any fumigant action. However, they did indicate that as a result of physiological reaction to metaldehyde, there is loss of water which continues even under 100 per cent Relative Humidity.

As regards the conditions under which metaldehyde would be able to affect higher mortality, Lange and MacLeod (1941) stated that more snails and slugs get killed in unshaded areas because of the killing action of sun. Woglum (1943) and Persing (1945) reported that substantial percentage of *Helix aspersa* recovered from metaldehyde poisoning 'during cloudy weather' or in the shade. Although these authors have not referred to the killing action of metaldehyde-poisoned snails but in substance they mean the same as Lange and MacLeod (1941). Similarly Corbett and Pagden (1941), without referring to sun, state that 'less meta is needed' for *A. fulica* when the bait is placed in 'exposed places'. In 1952, Lange and Sciaroni stated that 'in protected situations, metaldehyde does not seem to be an active enough poison to give high mortality'.

Under high relative humidity conditions Lange and MacLeod (1941) obtained greater kill but Stringer (1946) reported less favourable results because of greater recovery of slugs. Thomas (1948) concluded that the maximum kill with metaldehyde can be obtained during wet warm nights followed by dry warm days. Cragg and Vincent (1952) cleared the confusion. They explained that metaldehyde causes immobilisation, stoppage of feeding and secretion of mucus which results in loss of water from the body. Under optimum conditions of high humidity and low temperature, snails and slugs may recover; on the other hand when the humidity is low, there is greater loss of water and death occurs. Higher temperature produces greater activity of snails and higher incidence of death because of increased chances of coming in contact with metaldehyde and naturally greater uptake of the poison. However, increased activity of snails depends not only upon higher temperature, but upon high humidity also. Based on these informations, Moreton (1953) suggested that during dry weather lower concentration of metaldehyde may be used effectively. To stop the recovery of snails from metaldehyde poisoning, United States Bureau of Entomology and Plant Quarantine recorded in 1953 that the metaldehyde bait be fortified with calcium arsenate (Mead, 1961). However, at least in India it cannot be done because its use has been banned in this country because of its toxicity to warm blooded animals. Mead (1961) rightly concludes that variation in the results with metaldehyde are due to multiple and variable climatic conditions prevailing at a particular time.

The author would like to refer to two of his observations at Port Blair during the course of his work on the giant African snail.

At Port Blair, after observing the movement of the snail after coming out of the hiding places after dusk, 5 per cent metaldehyde pellets were spread all over the pathways for three days; there was no mortality; on the 4th day around 6 a.m., I got a telephone call from the then Horticultural Inspector saying that many snails were seen neither feeding nor moving-almost dead. I rushed to the Horticultural Garden and was delighted to see that the same pellets which had not killed any snail for the past three days led to death of a large number of snails on the 4th day. The reason : it had rained during the night and the sky was clear in the morning. During the scorching sun of the day, the metaldehyde-poisoned snails turned black and died. This shows very clearly that for metaldehyde killing high humidity followed by clear sky is necessary. Thomas (1948) also reported that maximum kill can be obtained when metaldehyde is used during wet and warm nights.

The second observation relates to the recovery of snails from metaldehyde poisoning. Once 5 per cent metaldehyde pellets were spread in the kitchen gardens of some staff members residing near Guest House No. 2. Next morning, a large number of snails were observed immobilised. It continued to rain almost continuously for 3-4 days and nights and the number of immobilised snails decreased. Even those which continued to remain immobilised probably due to higher uptake of metaldehyde took 3-4 days to die because there was no scorching sun during those 3-4 days. This is in conformity with the observations of Woglum (1943) and Persing (1945) as they report that 'substantial percentage of *Helix aspersa* (a land snail) will recover from metaldehyde poisoning during cloudy weather or in the shade" i.e., not exposed to scorching sun. Lange and MacLeod (1941) also observed that in "unshaded areas there was greater kill" because of the killing action of sun on the "stunned individuals". Lange and Sciaroni (1952) also stated that metaldehyde does not give "high mortalities" of snails in "protected situations".

Olson (1973) screened 107 chemicals (46 carbamates, 23 organophosphates, 19 salicylanilides, 6 tin compounds and 13 miscellaneous compounds), as molluscicide. Out of these, he got good results with 48 as bait, and held field trials also with 40 (21 carbamates, 9 organophosphates, 2 salicylanilides, 5 tin compounds and 3 miscellaneous compounds). He came to the conclusion that metaldehyde has been and remains the standard molluscicide used against *A. fulica* although it is not ovicidal in contrast to some carbamate compounds but carbamates have phytotoxic action on popular garden plants.

In Russia, Gorokhov (1971, 1972 a, b); Gorokhov and Aliev (1971) and Gorokhov and Krylov (1972) tested more than 200 molluscicides under different conditions against endemic and introduced land snails and slugs and 'metaldehyde emerged as the one preferred' (Mead, 1979). They used metaldehyde as spray and in bait form. A spray of 4-8 kg of metaldehyde in 600 litres of water/ha gave 98.7 per cent kill whereas 5 per cent bait as granule scattered @ 30-40 kg./ha gave only 75 per cent mortality. However, they preferred the bait because it is less toxic than spray, to fauna and flora.

Barnes and Weil (1942) tried a number of carriers for formulating a more effective bait but none proved better than the 'old standby wheat bran'. To add durability to the bait, Altson (1950) mixed cement and lime with metaldehyde to get hard brick-like bait. He found that the addition of rice bran to the bait made it more attractive to snails and slugs. Metaldehyde, lime, rice bran and cement were mixed in the ratio of 1 : 2 : 6 : 6. In recent years also, in New Hebrides, a mixture of metaldehyde and cement has been advocated to prolong 'life of the bait' (Mead, 1979). FitzGerald (1947) and Townes (1946) want metaldehyde to be used as such — as pellet, or dust without mixing it with anything else. However, this is not practicable as it is more expensive and less effective as evidenced by Speyer's conclusions (Speyer, 1954).

To increase the spreading effect (i) powdered metaldehyde with sand (Jary, 1939) and (ii) one per cent aqueous suspension spray (Alicata, 1950) have been used. Both these gave very good results. Cragg and Vincent's (1952) and Moreton's (1953) experimentations against slugs showed that use of metaldehyde as dust and spray lose their effectiveness "within a few days" hence the treatment has to be repeated periodically. However, Blowers (1954) states that a recently released commercial suspension of metaldehyde, leaves a semi-persistent coating of metaldehyde "sufficient . . . for many weeks". He, however, has not given any experimental data in support of his statement (Mead, 1961). On the Hawaiian Islands, metaldehyde dusts and sprays have been used against *A. fulica* and the spray was found to give better results (Weber, 1954).

Lewis and La Follette (1942a,b) concluded that under certain conditions, metaldehyde is repellent to *H. aspersa*. Lange and Sciaroni (1952) have stated that high concentration of metaldehyde becomes repellent. It was observed that metaldehyde bait, only up to 6 per cent concentration is effective and concentration higher than this becomes repellent (Anonymous, 1970).

Mead's recent correspondence from India, Mauritius, the Seychelles, Sarawak, New Guinea, Hong Kong, Taiwan, Yap and Tahiti, clearly indicates that "metaldehyde remains virtually the sole molluscicide" being used against *A. fulica*. The most common mode of its use is in the form of bait.

In India, different formulations of metaldehyde have been used against the giant African snail. Nair *et al.*, (1968) used metaldehyde dust and suspension. In the laboratory, one per cent metaldehyde suspension gave 100 per cent kill in 2 days whereas in the field cage 1.0 per cent gave the same result in 4 days. However, in the field test, one per cent suspension gave 93.3 per cent kill, only one day after the treatment; 2 per cent dust gave 100 per cent kill in 2 days in the laboratory and in 5 days in the field cage (Table 7.5). One per cent metaldehyde suspension spray "persisted in the field at a toxic level for 4 days".

TABLE 7.5: Laboratory and field-cage trials with metaldehyde against *A. fulica* in Kerala (After Nair *et al.*, 1968)

Treatment	Percentage mortality after (days)				
	1	2	3	4	5
Laboratory trials					
2% metaldehyde dust	64	100	—	—	—
4% metaldehyde suspension	92	100	—	—	—
2% "	84	100	—	—	—
1% "	80	100	—	—	—
Control (Treated with 1% Teepol)	0	0	0	0	0
Field cage test at Vellayani					
2% metaldehyde dust	0	16	48	76	100
1% metaldehyde suspension	13.2	53.3	83.3	100	—
Control (untreated)	0	0	0	0	0
Field test at Palghat					
1% metaldehyde suspension	93.3				
Control (untreated)	0	0	0	0	0

TABLE 7.6: Mortality of *A. fulica* exposed to surface sprayed with one per cent metaldehyde suspension (After Nair *et al.*, 1968)

(Ten snails were used on each occasion)

Interval after spraying	Percentage mortality after (days)						
	1	2	3	4	5	6	7
3 hours	30	100					
2 days	10	90	100				
4 days	10	50	50	50	50	80	100
6 days	10	30	70	70	70	70	70
8 days	0	0	0	0	0	0	0

Srivastava *et al.*, (1968) held field trials with 3 and 5 per cent metaldehyde pellets as bait against *A. fulica* at Port Blair (Andamans) and found 5 per cent bait to be better than 3 per cent (Table 7.7) Bhardwaj (1972) also has reported 5 per cent metaldehyde to be bett than 3 per cent against *A. fulica* (Table 7.3). Hoda (1971) has found 3 per cent metaldehyde pellets prepared with bran as carrier to be "highly effective" against snails in the garden. Srivastava and Abbas (1973) treated seedlings of papaya, cashewnut, lemon, orange, guava, sapota, emblic (*Emblica officinalis*) tamarind, tomato, brinjal, chillies, cotton, marigold, bougainvillea, arecanut, coconut, coffee etc., with 5 per cent metaldehyde dust. The snails (*A. fulica*) which approached the seedlings were trapped in the metaldehyde dust lying on the soil surface and could not damage the seedlings. The seedlings were observed for eight days and there was no phytotoxic effect. This observation is in conformity with the observations of Jefferson (1952) and Moreton (1953) who found the dusts and sprays of metaldehyde to have no phytotoxic effect even on such delicate plants as orchids. Thus, before transporting the seedlings from one place/island to another, these should be treated with 5 per cent metaldehyde dust so that if any snails happen to be carried with the seedlings/saplings they would die. This can be adopted as a domestic quarantine measure.

Saxena and Dubey (1970) working at Tezpur (Assam) on *A. fulica*, in addition to other chemicals, used metaldehyde bait also against the snail (Table 7.4). In bait A, they used "herbaceous base" and in bait B they used "some other base". They observed 96 and 98 per cent mortality after six hours of treatment. The percentage of metaldehyde used has not been mentioned but the quantity has been stated to be the same. The reason for the difference is attributed by the authors to the difference in the base used. These authors are of the view that metaldehyde is "often ineffective under moist conditions and tends to leave undesirable residues on food crops". Till date, as already stated earlier, metaldehyde has emerged as the best and safest chemical for the control of giant African snail. Bait formulation is most commonly recommended so the question of residue on food crops does not arise. Nobody who has handled the snails for some years will ever agree with the comments of these authors. Moreover, Nair *et al.*, (1968) have stated that metaldehyde suspension "persisted in the field at a toxic level for four days".

Before using the metaldehyde pellets, it is better to observe for 2-3 days the hiding places from where the snails start crawling out at dusk. Their pathways also should be observed. Later the pellets should be spread all over the pathways so that before reaching the food plants they get entrapped because of the pellets, become immobilised and finally die. Snails also hide during day time under garbage. In such

TABLE 7.7: Field trial with 3 per cent and 5 per cent metaldehyde pellets as baits against *A. fulica* at Port Blair (Srivastava *et al.*, 1968)

Host plants, plot/bed, size, No. of plants etc.	Date of bating	Observation			Remarks
		Date	Mortality number		
			with 3% metaldehyde	with 5% metaldehyde	
1	2	3	4	5	6
1. Papaya Plot size 20×30 feet No. of plants/plot 30-35 Height of plant 1-2 feet.	15.6.67	16.6.67	5	31	There was more or less continuous downpour throughout this period
		17.6.67	0	47	
		18.6.67	0	19	
2. Miscellaneous	5.6.67	6.6.67	5	40	
	16.6.67	20.6.67	3	28	
3. Portulaca	16.6.67	17.6.67	Not tested	13	
		18.6.67	"	17	
		19.6.67	"	6	
			"	6	

TABLE 7.7 (*Contd.*)

1	2	3	4	5			6
4. Zinnia		20.6.67	not tested	Side bed 1	Middle bed 2	Side bed 3	The middle bed of *Zinnia* was completely and the side beds partly obstructed by the main building of Horticultural Station, Haddo, Port Blair.
(a) Beds at lower level (12×12 feet)	16.6.67	17.6.67	"	4	2	3	
		18.6.67	"	5	1	4	
		19.6.67	"	10	0	5	
		20.6.67	"	20	5	10	
		25.6.67	"	10	1	4	
		26.6.67	"	15	0	4	
Zinnia (b) Beds at higher level circular (dia. 6 feet)	16.6.67	17.6.67	"	1	0	0	
		18.6.67	"	0	0	0	
		19.6.67	"	0	0	3	
		20.6.67	"	0	0	3	
5. Lady's finger Height of plants 1 feet — 1.5 feet Plot size 20×30 feet	24.6.67	25.6.67	"	80			
		26.6.67	"	110			

cases if metaldehyde pellets are spread all around the garbage, when they start moving out at dusk, they will get entrapped. If some particular seedling/sapling is to be protected, metaldehyde pellets can be spread all around evenly and very close together so that before reaching the particular seedling or sapling, they get immobilised by the pellets and finally die.

As regards the safety in the use of metaldehyde as a molluscicide, doubts started being expressed in its use soon after it started being used as a molluscicide. As early as 1937, Hudson raised an alarm by reporting the death of two birds, a black bird and a starling within about 30 minutes of feeding on metaldehyde bait. Shewell-Cooper (1938) reported the death of dogs by "eating meta" which was apparently placed for killing slugs. Lewis *et al.,* (1939) went a step further and presented a "thoroughly convincing case of death of a child from metaldehyde poisoning". Jary (1939) also suggested to be careful in using metaldehyde and stated that metaldehyde "might be harmful to earthworms". All this created an impression that the use of metaldehyde was not safe.

However, Cameron (1939) cleared these erroneous statements by feeding two pounds of 1 : 50 metaldehyde : bran mixture to two chickens in Ceylon (Sri Lanka) alongwith their regular feed for four weeks without recording any harmful effect on chickens even though this amount was much more than what they could have picked under the usual field conditions. Later Fernando (1952) also declared it safe for dogs, cats and poultry. Peterson (1957b) also states that spraying with solutions of wettable metaldehyde @ of 5 pounds per acre proved "very effective and safe to vegetation, animals and humans". The author, used metaldehyde pellets well over sixteen years for the management of *Achatina* on the Andamans in the field and did not get a single report of any accident. On the contrary, the metaldehyde pellets were in great demand by those keeping cows, poultry, ducks and pets like dogs and cats. Metaldehyde pellets (5%) were powdered, rubbed on the body of chicken fed to chicken and later spread over moist soil, and the chicken kept over it under a basket, yet no ill effect was observed. The predatory millipede *Orthomorpha* sp. the hermit crab and earthworms, all were found to be unaffected by metaldehyde. The millipede was observed nibbling the pellets and the earthworms crawling over these without being even immobilised.

Thus taking everything, including efficacy and safety to the fauna and flora into consideration, metaldehyde is the best chemical used for the management of *A. fulica*. This is the most widely used chemical in the *Achatina*-infested areas. While Olson (1973) considers it to be a "standard" molluscicide used against *A. fulica,* Mead (1979) states that Metaldehyde remains "virtually the sole molluscicide being used to control the giant African snail" mainly as bait in most of the countries.

Legislative Approach

From the very beginning when *A. fulica* started spreading from its original home land, East Africa, there have been appeals to local people to collect and destroy the specimens to prevent their spread into uninfested areas. Some suggested the inspection of all plants before allowing their entry into a country.

"Forcible and effective" quarantine regulations were enacted in Japan, against *A. fulica,* in 1936, but the supposed medicinal properties of *A. fulica* helped in its spread to newer areas (Mead, 1961).

In Hawaii, two-year old populations of *A. fulica* were discovered in Maui and Oahu in 1938 (Mead, 1949). Since then every effort is being made through general control measures and quarantine measures to eradicate this pest from the country.

Quarantine measures can be classified into (i) Domestic quarantine and (ii) International quarantine.

(i) *Domestic Quarantine*

Every country has its own rules and regulations for avoiding the spread of pests and diseases from one part of the country to another, but the success of domestic quarantine depends upon how strictly it is followed. In India, at least in case of *Achatina,* no heed appears to have been paid to the domestic quarantine measures as is evident from the reports of its spread to newer locations year after year.

(ii) *International Quarantine*

Under this, carrying of plant materials infested with pests and diseases from one country to another is prohibited. These legislations in effect led to an indefinite delay in the establishment of giant African snail in areas climatically suitable for their establishment. This delaying factor 'continues to be effective' in preventing the establishment of *A. fulica* in California. In Hawaii due to strong quarantine regulations enacted in 1938, the spread of this snail to major islands has been slowed down by at least 20 years (Mead, 1979). Australia is well known for implementing the quarantine regulations strictly. But in spite of this, the snail was located for the first time on the north-east coast of Queensland (Australia) (Coleman, 1977). According to Mead (1979) in spite of the quarantines, the snails are continuing to spread faster than more efficacious control measures are being discovered. He is of the view that the quarantine regulations should continue to be reviewed. He very rightly points out that "vast majority of personnel enforcing quarantine regulations against snails are entomologists with little or no training in malacology. The speed of the spreading of this snail in the world is faster than the speed with which researches on its control are being done (Mead, 1979).

Like other countries, India too has its own quarantine regulations. The Destructive Insects and Pests Act was passed in 1914 which has been amended from time to time. This act prohibits the import/export to/from/any part of India, of material likely to cause infection/ infestation to any crop. To discharge the international quarantine obligations, the Government of India have framed certain rules and regulations. Domestic quarantine is enforced through a series of notifications published in the Gazette of India, which restrict the movement of certain articles which are likely to help in the dispersal of notified pests and diseases (Wadhi, 1982).

In India, as far as *Achatina* is concerned, its introduction was not accidental but deliberate, bypassing international quarantine rules and regulations. It was introduced in Calcutta in 1847 and is gradually reaching different states of the country and getting established. It is becoming a problem for several economically important crops. As far as the islands of Andaman and Nicobar group are concerned, according to the old residents of these islands, it was introduced during the Second World War by the Japanese during the period of their occupation. This introduction of the giant African snail both on the main land of India as well as on its offshore islands was done by the then rulers.

If the domestic quarantine regulations were strictly followed as is being done in the case of the golden nematode of potato, further spread to newer areas can be checked. While the author was working on this snail (1965-1980) there was no report of its presence in Uttar Pradesh; however, in a recent survey conducted by Raut and Ghose (1984) it has been located in Moradabad (Uttar Pradesh). Similarly this snail was found only in a few islands of Andaman and Nicobar group but now it is reported to have reached many of the inhabited islands. If the seedlings/saplings were checked/treated with some suitable chemical, the snails could have got killed at the place from where these were transferred to other places/islands (Srivastava and Abbas, 1973).

In addition to legal aspects, there is a moral aspect also under which it becomes the responsibility of every citizen of the country to see that it does not spread further so that if the problem is not solved atleast it may not become more complicated.

Integrated Approach

It is evident that no single method is sufficient to bring down population of economically the most important land snail pest *Achatina fulica.* The only practical method to deal with this pest is the integrated approach in which all such methods are to be used simultaneously which are not antagonistic but complementary to each other. To quote an example, where metaldehyde pellets are used to kill *Achatina,* predatory snails cannot be used as biological control agents, for they also would get killed by the molluscicide which kills *Achatina.* In the integrated approach, snails which do not get killed by one method may get killed by the other methods.

As a result of 16 years of continuous work on *Achatina* by the Indian Agricultural Research Institute, New Delhi and its substation at Port Blair, an integrated management for *A. fulica* has been evolved (Fig. 7.12) firstly, by conducting laboratory tests and later by conducting field trials on certain islands of Andaman and Nicobar group. The integrated management of the giant African snail does not aim at its eradication but aims at reducing the population below the economic injury level; the control measures have to be undertaken almost all the year round i.e., during the season of its activity as well as during the season of its inactivity i.e., during hibernation/aestivation.

This approach includes collection and destruction of snails and their egg masses, location and destruction of day-hide outs of the snail during the season of activity wherever, feasible and covering the pathways of snails with 5 per cent metaldehyde pellets. In Andaman and Nicobar islands and other places where hermit crabs of genus *Coenobita* are available, collection and release, in the snail-infested locations where they may locate and feed upon the snails (Srivastava, 1976; Srivastava and Abbas, 1983) and protecting the firefly *Lamprophorus tenebrosus* larvae and collection and release of these in snail-infested locations may be done. During the season of inactivity, location and destruction of the hibernating/aestivating pockets, along with the snails hiding therein, also forms an integral part of integrated management of this pest. This is very important, for, during hibernation/aestivation larger number of snails remain in smaller areas; if off-season pockets are located and destroyed, comparatively very little number will be left behind to become active on the onset of monsoon and naturally the damage by them will be much less.

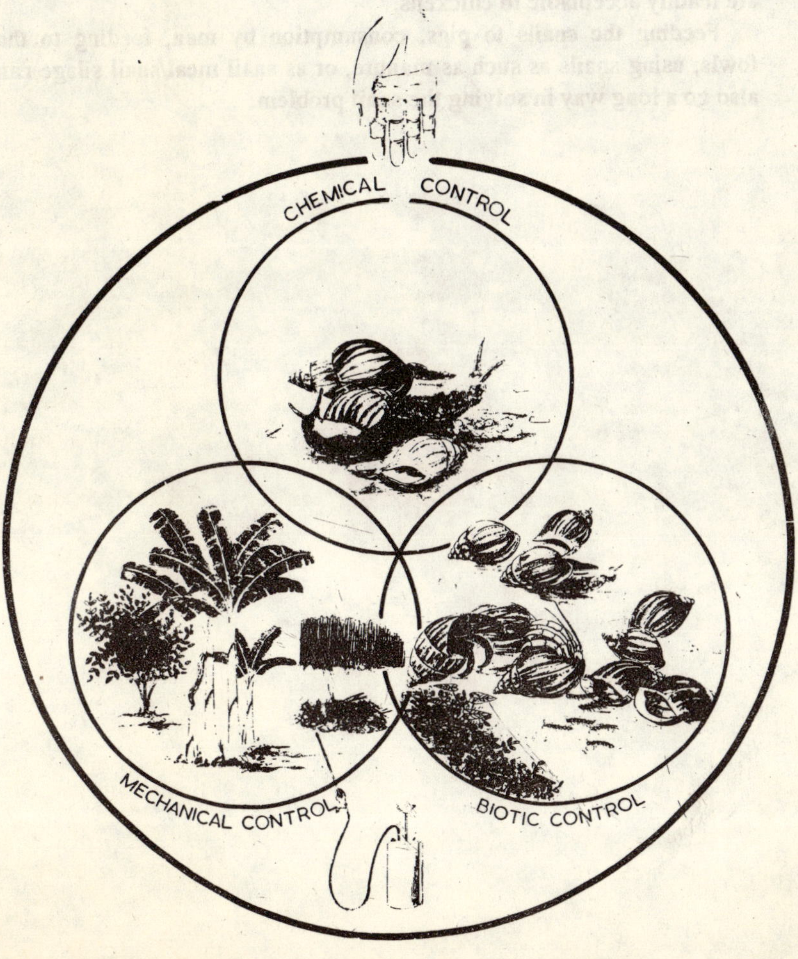

FIG. 7.12: Integrated approach for the management of the giant African snail

Wherever possible non-food plants of *Achatina* should be grown. Keeping ducks serves double purpose; while one gets eggs from these birds, they keep the kitchen gardens free from eggs, infants and juveniles and these are not allowed to become reproducers. The adults not killed by management practices "become non-reproductive or sterile with time" (Olson, 1973). This will, however, hold good, only if migration from adjoining areas could be reduced to the minimum. Big-sized crushed snails can be fed raw to ducks. Boiled crushed snails are readily acceptable to chickens.

Feeding the snails to pigs, consumption by man, feeding to the fowls, using snails as such as manure, or as snail meal/snail silage can also go a long way in solving the snail problem.

8

FUTURE OVERVIEW

From the aforegoing account of *Achatina fulica* Bowdich, it is clear that at least in India very little attention has been paid to the applied aspects of land snails including *A. fulica*.

Since the snails remain hiding during day time even under dry/drying fallen leaves of plants, some people have a wrong notion that they are not pests and feed only on fallen leaves of plants. According to Schreurs (1963) green parts of plants are essential in the diet of *A. fulica* for the first six months of development. This indicates that this snail eats more live plants up to the age of six months. In spite of all the efforts of scientists and all the vigil of quarantine staff, the giant African snail continues to spread from country to country and place to place. Mead (1979) states that "the snails are continuing to spread faster than more efficacious control measures are being discovered".

So far as India is concerned surveys have revealed that after its introduction in Calcutta more than 140 years ago, it has established and spread to many states of the country; on the Andamans it is reported to have been introduced by the Japanese during the Second World War and since then it has been spreading from place to place and island to island unchecked. During the course of over sixteen years of the author's work on *Achatina*, he did not observe enforcement of any thing like domestic quarantine any where in the country so far as *Achatina* is concerned. The spreading of this snail is not being taken seriously. The snail problem of the Andamans was given to the Indian Agricultural Research Institute, New Delhi for tackling by the Government of India in 1964. Later it became a five year plan project and finally a non-plan project. The Institute started working on the project in right earnest with a small unit — some of the staff posted at its substation at Port Blair and some at Delhi. Laboratory researches were carried out mainly at Delhi and field trials were conducted on South Andaman. In late seventies, with the establishment of a new

Institute of the Indian Council of Agricultural Research at Port Blair, the substation of the Indian Agricultural Research Institute was merged with it. Since then not much has been done regarding the snail at Port Blair, neither by the new Institute nor by the Department of Agriculture of Andaman and Nicobar Administration. Even the recommendations of the project on *Achatina* on which the author and his co-workers worked for over one and a half decades are not being implemented adequately. Staff of the erstwhile substation got transferred to the mainland of India and now there is no one at Port Blair working solely on snails. The Indian Agricultural Research Institute also closed down the snail project with effect from 31-12-80 and at present no work is being done by it on snails at Delhi.

There are 41 Research Institutes of the Indian Council of Agricultural Research and 26 Agricultural Universities, spread all over the country yet hardly any work on snails is being done any where. Many of the Institutes are not committed to any particular crop/animal and work on *Achatina* and other land snail pests can be taken up easily, at least by those institutes/universities which are located in snail-infested states such as Assam, West Bengal, Bihar, Orissa, Kerala, Tamil Nadu and Maharashtra. The only work on land snails which the author is aware of, is being done on the land snail *Cryptozona belangiri* at the College of Agriculture, Kolhapur, a constituent college of Mahatma Phule Krishi Vidyapeeth, Rahuri (Maharashtra) under a small project financed by the Indian Council of Agricultural Research. The Council should advise and sanction projects to other Agricultural Universities also for taking up work on land snails in general and *Achatina* in particular. Some work is also being done at the Konkan Krishi Vidyapeeth, Dapoli (Maharashtra). The University of Calcutta which is not an agricultural university, is a pioneer in the field of research on land snails especially *Achatina* for over three decades. Although their work is mainly on the fundamental aspects which are also important and help in devising control programmes, they have been doing some work on applied aspects also. It is high time that the Central Institute of Agricultural Research at Port Blair pays more attention to *A. fulica* as it has already reached a number of islands. The Andaman Administration should also take steps to enforce quarantine rules and regulations and treat seedlings/saplings with molluscicides before shifting from one place/island to another.

Before finalising the programme for the management of the giant

African snail, certain points have to be taken into consideration.

Often eradication of this snail is talked of. To think of eradication of an animal, gifted by nature with so many ways and means to protect itself, is unrealistic. The scientists on the Hawaiian islands, with planty of resources and funds at their disposal have not yet been able to eradicate this pest from there, how can we think of eradicating it with the meagre resources at our disposal and with the non-enforcement of domestic quarantine at least regarding *Achatina*? The realistic and practical approach to this problem would be to take such measures which may bring down the population below the economic injury level.

Work on the snail management should be regulaı and continuous. Stray work here and there as has been done in Assam, Bihar, Orissa and Kerala is not going to solve the snail problem.

Work on the snail pest management is generally carried out for a few months only during the season of activity. It should be done during the season of inactivity also by locating and destroying the hibernating/aestivating pockets. So far the author has not come across any publication from any where on the location and destruction of hibernating/aestivating pockets along with the snails hiding therein, except of the author and his co-workers.

To make the snail management programme a real success the major requirements are as follows:

The method (1) should be effective and should be able to kill more than 50 per cent of the pest, (2) should cause least pollution of the environment, (3) should not be harmful to warm blooded animals and useful soil fauna and (4) should not be phytotoxic to the crops for whose protection control measures have to be undertaken. Finally it should be undertaken under the direct supervision/guidance of an expert and should not be left to the inexperienced/unskilled staff. In the case of this snail, quarantine, both international and domestic has to play very significant role.

The requirements given above for the snail pest management are ideal; no single method meets all the requirements hence none of the methods can be ignored.

Collection and destruction of the snails and their egg masses is quite easy. Olson (1973) has concluded that "baiting and hand-collecting are equally effective and are complementary". However, proper disposal of the collected snails is very important. Either these should be buried in the fields far away from habitations, or without shells in the agricultural fields to be utilised as manure. If the

Government Departments can pay in cash against a fixed weight of snails collected it would give them incentive. Egg collection can be done during day time also. As regards snail collection, more snails can be collected at any time between dusk and dawn. This approach to the snail problem poses no threat to the environment. However, it does involve several man-hours and man-days.

Destruction of hiding places of snails both during the season of activity and during hibernation/aestivation is very effective in reducing the snail population, especially the destruction of aestivating/hibernating pockets. Location and destruction of hibernating/aestivating pockets is very effective because larger number of snails hibernate/aestivate in much smaller areas. Moreover, this can be done during day time. This method is safe also from all angles.

Crops can also be saved by growing favourite food plants of the snail as a trap crop all around the crop required to be protected. This approach will keep the snails busy in the trap crop and may not be able to reach the main crop.

In areas where snail population is very high, a practical approach can be to grow those crops which the snail does not feed upon, than to spend money and time in controlling the snails. Coconut, tea, sugarcane, pineapple, paddy, wheat and bittergourd are some of the crops which the snail does not feed upon. There are some reports of slight damage to maize crop especially at the seedling stage. If the seedlings could be protected, there should be no problem for growing the maize crop. Choice of the non-food plant crops to be grown in a particular area will depend on the climatic conditions of the area concerned.

Biological control methods are safe and do not cause any pollution and pose no threat to the life of warm-blooded animals. These involve the use of predators, parasites and pathogens. Predators can locate the prey and feed upon it, but these are non-specific, and, in addition to the snail, these prey upon insects etc. also. So far the main predators used for the snail control have been molluscs, insects and crustaceans. Molluscan predators have the draw back that these cannot be used along with chemical molluscicides as the molluscicides which kill the pest snails also kill the predator snails. In addition, the biotic potential of the predatory snails is much lower than that of *A. fulica*.

Amongst the non-snail predators, the main predators used against the giant African snail are the beetles especially the Indian glow worm, *Lamprophorus tenebrosus* and hermit crabs of the genus *Caoenobita*. The glow worm larvae were collected from Sri Lanka several times and released in snail infested locations mainly on the Hawaiian islands by different workers but the overall performance has not been satisfactory. Only in some cases, some recoveries were made. Their inability to reduce the snail population effectively may be either because of their not being in adequate numbers or because of their being predated upon by other predators. What is required is to develop the technique of mass rearing this predator in the laboratory so that it may be released in large numbers and the snail-infested locations may be inundated with this predator. Another point worthy of being taken notice of is that in spite of the Indian glow worm being present in large numbers in Sri Lanka, Andamans and Guwahati (Assam) *Achatina* continues to be a problem even in these areas.

Amongst the crustacean predators, hermit crabs of genus *Coenobita* are effective predators. Efforts to establish these in snail-infested locations quite away from sea-shore have not met with success. They do remain alive for a few months in such areas and reduce the snail population but do not establish there. Thus, it is feasible to collect and release these in snail infested locations only on islands or in areas not far away from sea shore. Their release in inland areas like Pusa (Bihar) and Guwahati (Assam) is not feasible. So far, the technique to breed these under simulated sea-shore conditions have failed. Further efforts are required for breeding it under simulated sea-shore conditions.

There are records of snail disease from India and abroad. Spraying aqueous extract of diseased snails over the field population of the snail has succeeded in affecting mortality. Disease has been reported to be the probable cause of population decline in some areas. There is need to develop the technique for preparing mass culture of the causative organism so that it could be used for the control of field population.

In spite of the biological control approach being a very safe method from practically all angles, a lot has yet to be done to make it a practical and feasible proposition. Thus, biological control methods in Economic Malacology are still not so well developed as to be undertaken successfully to reduce the snail population in the field. Amongst the chemicals, it is now well established that metaldehyde as bait in the form of pellets is the best, from the point of view of efficiency and safety to other animals and plants. In India, two firms are registered with the Directorate of Plant Protection, New Delhi for formulating it, one in Bombay and the other in Udaipur but for the past several years, they have not been formulating it obviously because it is not very profitable as it can be used only against snails, whereas other pesticides can be used against may pests which occur practically all over the country hence would be in great demand. In public interest, the Government of India should allot this work to some Government of India undertakings.

Keeping the ducks protect at least the kitchen gardens from the snail menace. They feed upon the eggs and young ones up to 3 cm shell-size; in kitchen gardens snail collection and their destruction is not difficult and thus, the kitchen gardens can be kept free from adult snails, their eggs and young ones. However, vigil will have to be kept to avoid migration from the neighbouring areas. Such migrations can be reduced effectively if people in all the kitchen gardens in a particular area keep ducks and collect and destroy adult snails.

Use of the snail as a part of animal feed, poultry feed and human diet can go a long way in solving the problem of snail menace. Those who keep pigs can collect snails and feed to their pigs and thus save some money spent on feeding the pigs. Snail meal can be easily prepared and form a part of poultry feed.

As regards human consumption, Mead (1982) has very rightly said that it is pity that in several snail-infested areas children are seen suffering from malnutrition although such a rich source of protein as *Achatina* which has 60 per cent protein (Table 6.3) in dry state is, abundant in such areas. When developed countries like France, Canada, and the U.S.A. can eat snails why can't the non-vegetarians in India and many other *Achatina*-infested countries eat it? Our neighbouring countries Thailand, Taiwan and Indonesia are processing and exporting *Achatina* to France, Canada and the U.S.A. Every effort should be made to process and export this snail. A beginning appears to have been made by the Marine Products Export Development Authority (MPEDA). Deputing some scientists to Thailand/Taiwan/ Indonesia to get training in processing the snail's meat, may also be considered so that the process of export is expedited.

A note of caution ! *Achatina* is not only a pest but a scavenger also because 75 per cent of its diet, consists of scavenger-type of food. Thus, while causing damage to crops it is also doing a service to humanity by consuming the dead and dying animal and plant tissues, which otherwise might have involved lot of man hours and man-days for disposal. Thus, they are maintaining some sort of balance in nature; if we remove *Achatina* from the scene completely, the balance in nature is likely to be upset. Thus, our effort should be to reduce its population to such a level that it does not remain economically harmful to crops and side by side balance in nature is not disturbed.

REFERENCES

Abbas, S.R. and Gautam, S.S.S., 1975. *The Veliger*, **17** (3) : 311-312.

Alicata, J.F., 1950. *Bull. Pacific Orchard Soc. Hawaii*, **8** (3) : 279-285.

Altson, R.A., 1950. *Rubber Res. Inst., Malaya, Report for the perioc Jan., 1941 to Aug., 1945*: 67.

Annandale, N., 1919. *Nature*, **104** : 412-413.

Annandale, N. and Prasad, B., 1920. *Rec. Indian Mus.*, **19** : 189-194.

Anonymous, 1970. *Int. Pest Control*, July/Aug., : 45.

Ash, L.R., 1967. Seminar on Helminthiases and Eosinophilic Meningitis, Noumea, New Caledonia, June 5-16, : 9 pp. Noumea, South Pacific Commission Working paper No. 13.

Atkins, W.R.G. and Lebour, M.V., 1923. *Roy. Dublin Soc. Sci., Proc., N.S.*, **17** (25-31) : 233-240.

Barnes, H.F. and Weil, J.W., 1942. *Ann. appl. Biol.*, **29** (1) : 56-68.

Beeley, F., 1938. *J. Rubber Res. Inst. Malaya*, **8** (2) : 130-39. Reprinted : *Trop. Agriculturist, Ceylon*, **90** : 247-254.

Behura, B.K., 1955. *J. Bombay nat. Hist. Soc.*, **53** (2) : 287-288.

Benson, W.H., 1858. *Jour. de Conchyl.*, **7** : 266-268.

Bequaert, J.C., 1950. *Bull. Mus. Comp. Zool.*, Harvard, **105** (1): 1-216.

Bhalla, O.P. and Pawar, A.D., 1977. *A survey of insect and non-insect pests of economic importance in Himachal Pradesh.* 80 pp. Monog. Dept. of Entomology Zoology, College of Agriculture, Chambaghat, Solan (H.P.).

Bhardwaj, A.K., 1972. *Indian J. Ent.*, **34** : 42-45.

Bhardwaj, A.K. and Thakur, J.R., 1973. *Curr. Sci*, **42** (21) : 769.

Bhat, P.K. and Shamanna, H.V., 1973. *J. Coffee Res.*, **3** (1) : 24-27.

Bhattacharjee, P.C., 1980. *Proc. 5th All India Congress Zool.*, Bhopal : 122.

Blanford, W.T. and Godwin-Austen, H.H., 1908. *Fauna of British India. Mollusca* : Taylor and Francis Red Lion Court, Fleet Street, London.

Blowers, J.W., 1954. *Orchid Rev.* London, **62** (730) : 56.

Boettger, C.R., 1951. *Braunschweigischen Wissenschaft. Gesell.*, **3** : 10-18.

Boycott, A.F., 1934. *J. Ecol.*, **22** : 1-38.

Brand, T.F. Von., 1931. *Zeit. Vergl. Physiol*, **14** : 200-264.

—, 1932. *Archivio. Zool. Italiano Torino*, **16** : 871-878.

Breckenridge, W.R. and Fallil, S.F.M., 1973. *Ceylon J. Sci.* (Bio. Sci.) **70** (2) : 85-118.

Brumpt, V., Audebaud, G., Klein, J.M., Jolly, M., Mazaud, R. and Goube, P., 1968. *Bull. Soc. Path. Exot.*, **61** : 444-462.

Buckhurst, A.S., 1947. *Fruit Grower* (London) **103** : 146, 149.

Burch, J.B., 1955. *Nautilus*, **69** : 34-46.

Butot, L.J.M., 1952. *Tropische Natuur*, **32** (2/3) : 56-58.

Cameron, A.E., 1939. *Jour. Ministry Agric.* Great Britain, **46** (5) : 454-462.

Chock, Q.C. and Nakao, H.S., 1951. Hawaii Board of Commissioners of Agric. and Forestry, Div. Ent. and Marketing, 4 pp.

Choy, R.B., 1963. *Philipp. J. For.*, **19** : 99-107.

Coleman, P.H., 1977. *N. Queensland Nat.*, **44** : 8.

Corbett, G.H., 1933. *Malayan agric. Jour.*, **21** (2) : 77-79.

Corbett, G.H. and Pagden, H.T., 1941. *Malayan Agric. Jour*, **39** (9) : 347-375.

Cragg, J.B. and Vincent, M.H., 1952. *Ann. Appl. Biol*, **39** (3) : 392-406.

Creswell, D.C. and Kompiang, I.P., 1979. *Proc. 3rd Australian Poultry and Stock Feed Convention*, 33-38 pp.

Crowell, 1967. *J. Econ. Ent.*, **60** : 1048-1050.

Davis, C.J., 1954. Invert. Consultants Comm. for the Pacific, Pac. Sci. Board., Nat. Res. Council. 24 pp. (Mimeographed).

—, 1958. *Proc. Hawaii. Entomol. Soc.*, **16** (3) : 356-358.

—, 1960a. *Proc. Hawaii. Entomol. Soc.*, **17** : 170.

—, 1960b. *Proc. Hawaii. Entomol. Soc.*, **17** : (2) : 244-248.

—, 1961a. *Proc. Hawaii. Entomol. Soc.*, **17** : (3) : 389-390

—, 1961b. *Proc. Hawaii. Entomol. Soc.*, **17** : 329.

—, 1961c. *Proc. Hawaii. Entomol. Soc.*, **17** : 325-326.

—, 1972 *Proc. Hawaii. Entomol. Soc.*, **21** : 187-190.

Davis, C.J. and Butler, G.O. Jr., 1964. *Proc. Hawaii. Entomol. Soc.*, **18** : 377-389.

Davis, C.J. and Chong, M., 1969. *Proc. Hawaii. Entomol. Soc.*, **20** : 317-322.

Davis, C.J. and Krauss, N.L.H., 1962. *Proc. Hawaii. Entomol. Soc.*, **18** (1) : 125-129.

——, 1964. *Proc. Hawaii. Entomol. Soc.*, **18** : 319-397.

——, 1966. *Proc. Hawaii. Entomol. Soc.*, **19** : 201-207.

Davis, T.A., 1987. *Environmental Awareness*, **10** (1) : 13-24.

Dean, R.E., 1950. *A Rep. Superintendent of Gardens for the period ended the 31st March 1950.* Hong Kong, 10 pp.

Dean, W.W., Mead, A.R. and Northey W.T., 1970. *J. Inverteb. Path.*, **16** : 346-351.

Deka, T.C., Bhattacharjee, P.C. and Bhattacharya, I.K., 1985. *In* : *Non-insect pests and predators* : 79-83. All India Scientific Writers' Society, New Delhi.

Dresner, 1958. *J. Econ. Ent.*, **51** : 390-391.

Dufo, H. 1840. *Ann. Sci. Nat.* 2nd Ser. (Zool.), **14** : 45-80. 116-221.

Dun, G.S., 1967. *Papua New Guinea Agric. J.*, **18** : 123-215.

Dupont, R., 1935. *Soc. Roy. des Arts et des Sci. de Maurice, Trans.*, Ser. C. 1933/34 (3) : 138.

Eisner, T., Hurst, J.J. and Meinwald, J., 1963 a. *Psyche, Camb.*, **70** : 94.

Eisner, H.E., Eisner, T. and Hurst, J.J., 1963 b. *Chemy. Ind.*, **3** : 124.

Esaki, T. and Takahashi, K., 1942. *Jour. Palan Trop. Biol. Sta.*, **4** (3) : 16-25.

Fernando, H.E., 1952. *Ceylon Dept. Agric. Propaganda Handbill.* 4 pp.

Fischer, P.H. and Huynh, C.C., 1966. *J. Conchyliol*, **105** : 149-150.

FitzGerald, L.D.E.F.V., 1947. *Rubber Res. Inst., Malaya A. Rep.* 1940 : 111.

Fromming, E., 1949. *Zeitsch. f. Hygienische Zool. U. Schadlings*, **37**: 320-325.

——, 1950. *Anz. f. Schadlingskunde*, **23** : 57-60.

Fromming, E. and Plate, H.P., 1952. *Zeitsch. f. Hygienische Zool. u. Schadlings*, **44** : 130-131.

Fromming, E. and Remschneider, R., 1952. *Anz. f. Schadlungskunde*, **25** : 186-189.

Fullaway, D.T., 1939. *Hawaii Board of Commissioners of Agric. and Forestry, Biennial Rep.*, 1937-38 : 24.

——, 1949. *Hawaii Board of Commissioners of Agric. and Forestry, Biennial Rep.*, 47-48 : 69-70.

Garnadi, P., 1951. *Hemera Zoa*, **58** : 279-301.

Ghose, K.C., 1959. *J. Bombay nat. Hist. Soc.*, **56** (2) : 183-187.

——, 1960. *Proc. zool. Soc.* Calcutta, **13** : 91-96.

——, 1962a. *J. Anim. Morph. Physiol.*, **9** (2) : 131-134.

——, 1962b. *Ann. Zool.*, **4** (3) : 25-37.

——, 1963a. *J. Bombay nat. Hist. Soc.*, **60** : 228-232.
——, 1963b. *J. Anim. Morph. Physiol.*, **10** (1) : 80-82.
——, 1963c. *Trans. Amer. Micros. Soc.*, **82** (2) : 149-167.
——, 1963d. *Proc. zool. Soc. Lond.*, **140** : 681-695.
——, 1964a. *Philipp. J. Sci.,*, **93** (2) : 219-229.
——, 1964b. *Proc. zool. Soc.* Calcutta, **17** : 159-167.
——, 1966. *The Veliger*, **9** (1) : 25-28.
Gimingham, C.T., 1940. *Ann. Appl. Biol.*, **27** (2) : 167-168.
Godwin - Austen, H.H., 1908. *Proc. Mala. Soc.* London, **8** : 146-147.
Gorokhov, V.V., 1971. *Mater. Nauch. Konf. Vses. Obshch. Gelmint* 1969-70, **23** : 89-92.
——, 1972a. : Problemy Parazitologii, Trudy VII Nauchnoi Konferentsii Parasilogov. Kiev, USSR Part 1: 223-225.
——, 1972b. *Mater Nauch Konf. Vses. Obshch. Gelmint.*, 1970/71, **24** : 40-43.
Gorokhov, V.V. and Aliev, N.A., 1971. *Byull. Vses. Inst. Gelmint.*, **5** : 15-19.
Gorokhov, V.V. and Krylov, A.I., 1972. *Byull. Vses. Inst. Gelmint.*, **9** : 31-36.
Gould, H.J. and Webley, D., 1972. *Plant Path.*, **21** : 77-82.
Green, E.E., 1910. *Trop. Agriculturist.*, **35** (2) : 120-121.
——, 1911. *Zoologist 4th Ser.*, **15** (170) : 41-45.
Gude, G.K., 1914. *Fauna of British India. Mollusca II.* Taylor & Francis, London.
Gupta, G.P., Abbas, S.R. and Srivastava, P.D., 1980. *Indian J. Ent.*, **42** (4) : 794-795.
Gupta, G.P., and Doharey, K.L., 1985. *In* : *Non-insect Pests and Predators*: 84-90. All India Scientific Writers' Society, New Delhi.
Gupta, G.P., Gautam, S.S.S., Abbas, S.R. and Srivastava, P.D., 1978. *The Veliger*, **21** (1) : 135-136.
Hall, A.H., 1932. *Nature*, **130** (3274) : 170.
Hatai, S., and Kato, G., 1943. *Jour. Palau Trop. Biol. Sta.*, **5** (2) : 1-19.
Hely, P.C., 1946. *Agric. Gaz. N.S. Wales*, **57** (5) : 239-240.
Hoda, M.Q., 1971. *Intensive Agric.*, **9** (7) : 9-10.
Hora, S.L. and Srinivasa Rao, H., 1927. *Rec. Indian Mus.*, **29** : 49-62.
Howes, N.H. and Wells, G.P., 1934a. *J. Exp. Biol.*, **11** : 327-343.
——, 1934b. *J. Exp. Biol.*, **11** : 344-351.
Hudson, R.N., 1937. *The Gardner's Chronicle*, 3rd Ser., **102** (2644) : 153.
Hunter, P.J. and Symonds, B.V., 1970. *Ann. Appl. Biol.*, **65** : 1-7.

Hutson, J.C., 1920. *Trop. Agriculturist*, **55** : 217-225.
Hyman, L.H., 1967. *The Invertebrates*, **6** : *Mollusca*, McGraw Hill Book Co., New York, London.
Jagtap, A.B., Mali, M.S. and Moholkar, P.R., 1989. *Pestology*, **13** (2) : 13-15.
Jarrett, V.H.C., 1923. *Singapore Naturalist*, **1** (2) : 73-76.
—, 1931. *Hong Kong Naturalist*, **2** (4) : 262-264.
—, 1932. *Hong Kong Naturalist*, **3** (1) : 70.
—, 1949. *The Times*, (London) No. 51482 : 5.
Jary, S.G., 1939. *Scientific Hort.*, **7** : 180-185.
Jary, S.G. and Austin, M.D., 1937. *J. South-eastern Agric. Coll.*, Wye, Kent., **40** : 183-186.
Jefferson, R.N., 1952. *Florists' Rev.*, **111** (2869) : 29-30, 123-127.
Johnson, W.H., Delanney, L.E., Williams, E.C. and Cole, T.A., 1969. *Principles of Zoology*, Holt, Rinehart and Winston, Inc., New York, USA.
Judge, F.D., 1969. *J. Econ. Ent.*, **62** : 1393-1397.
Kaburaki, T., 1940. *Proc. Sixth Pac. Sci. Cong.*, **4** : 229-230.
Kaestner, A., 1967. *Invertebrate Zoology*, Vol. 1, Inter Science Publishers, John Wiley & Sons, New York, London.
Kalshoven, L.G.E., 1950. In his : *De plagen van de cultuurgewassen in indonesie*, 2 Vols. 'S Gravenhage, Bandoeng, Van Hoeve, **1** : 46-59.
Kamada, T., 1933. *J. exp. Biol.*, **10** (1) : 75-78.
Karlin, E.J., 1961. *Nautilus*, **74** : 125-130.
Karlin, E.J. and Naegele, J.A., 1958. *N.Y. State Coll. of Agric., Cornell Univ. Ext. Bull.*, **1004** : 15 pp.
Kekauoha, W., 1966a. *Nautilus*, **80** (1) : 3-10.
—, 1966b. *Nautilus*, **80** (2) : 39-46.
Kompiang, I.P., 1979a. *Proc. Indonesian Animal Science Seminars*, Bogor, 26 Oct., 1979, Abstract.
—, 1979b. *Proc. 2nd Poultry Science and Industry Seminar*, P 3 T, : 165.
Kondo, Y., 1950a. The giant African Snail *(Achatina fulica)* in Palau, Pagan and Guam. Invert. Consultants Comm. for Micronesia, Pac. Sci. Board, Nat. Res. Council, 13 pp. (Mimeographed).
—, 1950b. Report on the *Achatina fulica* investigation on Palau, Pagan and Guam. Invert. Consultants Comm. for Micronesia, Pac. Sci. Board, Nat. Res. Council, 94 pp. (Typescript).

—, 1952. Report on carnivorous snail experiment on Agiguan island, primary and secondary *Achatina*-free areas on Rota and gigantism among *Achatina* on Guam. Invert. Consultants Comm. for Micronesia, Pac. Sci. Board, Nat. Res. Council, 50 pp. (Mimeographed).

—, 1964. *Nautilus,* **78** (1) : 6-15.

Krauss, N.L.H., 1964. *Nautilus,* **78** : 21-27.

Kulkarni, A.B. and Nagbhushanam, 1973. *Marathwada Univ. J. Sci. (Biol. Sci.),* **12** (5) : 77-92.

Landry, S.O. Jr., 1970. *Quart. Rev. Biol.,* **45** : 351-372.

Lange, W.H., 1950. *Pac. Sci.,* **4** (4) : 323-335.

Lange, W.H., Jr. and MacLeod, G.F., 1941. *J. Econ. Ent.,* **34** : 321-322.

Lange, W.H. Jr. and Sciaroni, R.H., 1952. *J. Econ. Ent.,* **45** (5) : 896-897.

Latif, S.M., 1933. *De Orchidee* (Bandoeng), **2** (7) : 178.

Laxminarayana, K.V. and Subba Rao, N.V., 1971. *The Andhra Agric. J.,* **18** (2) : 82.

Leefmans, S., 1933a. *De Orchidee* (Bandoeng), **2** (7) : 151-153.

—, 1933b. *De Orchidee* (Bandoeng), **2** (8) : 199-200.

—, 1933c. *Landbouw, Landbouwkundig Tijdschr. voor Nederlandsch-Indie,* **9** (6) : 289-298.

Leeuwen, D. van., 1932. *Hong Kong Naturalist,* **3** (1) : 71.

Lesson, R.P., 1830. In : *Louis I. Duperrey, Voyage autour du monde sur la Caquille, pendant les annee's,* 1822, 1823, 1824 and 1825, **2** (1) : 318-320.

Lewis, D.R., Madel, G.A. and Drury, J., 1939. *Brit. Med. Jour.,* **1939** (1) : 1283-1284.

Lewis, H.C. and La Follette, J.R., 1942a. *Calif. Citrograph,* **27** (5) : 122, 144.

—, 1942b. *J. Econ. Ent.,* **35** : 359-362.

Lim, B.L. and Onar-Ahmad, U.D., 1969. *Med. J. Malaya,* **23** : 208-213.

Lim, B.L., Ow-Yang, C.K. and Lie, K.J., 1965. *Amer. J. Trop. Med. Hyg.,* **14** : 610-617.

Lloyd, R.E., 1910. *An Introduction to Biology,* Longmans Green & Co., London.

Mandal, T.K. and Ghose, K.C., 1970. *Indian J. exp. Biol.,* **8** : 332-333.

Manna, B., 1976. Possibility of the use of organophosphorus chemicals, Diazinon and Fenitrothion as gut poison in the control of *Achatina fulica*. Ph.D. Thesis, University of Calcutta, Calcutta.

Manna, B. and Ghose, K.C., 1972. *Indian J. exp. Biol.*, **10** : 461-463.

Mead, A.R., 1949. *Atlantic*, **184** (2) : 38-42.

—, 1950a. *Bull. Mus. Comp. Zool.* Harvard, **105** : 218-242.

—, 1950b. The giant African snail problem (*Achatina fulica*) in Micronesia. Final report. Invert. Consultants Comm. for Micronesia, Pac. Sci. Board, Nat. Res. Council, 55 pp. (Typescript).

—, 1950c. The problem of the giant African snail. Final report, Invert. Consultants Comm. for Micronesia Pac. Sci. Board, Nat. Res. Council, 30 pp. (Mimeographed).

—, 1955. *Ceylon Forester, N.S.*, **2** (1) : 47-50.

—, 1958a. *Amer. Malacol. Union, A. Rep.* : 37.

—, 1958b. *Amer. Assoc. Adv. Sci., South-western and Rocky Mountain Div., Program 34th A. mtg* : 22.

—, 1959a. *Proc. Hawaii. Entomol. Soc.*, **17** (1) : 85-86.

—, 1959b. *Jour. Colo. Wyo. Acad. Sci.*, **4** (11) : 51-52.

—, 1961. *The giant African snail : A problem in economic Malacology*, 257pp. The University of Chicago Press, Chicago, USA.

—, 1963a. *Malacologia*, **1** : 305-311.

—, 1963b. *Amer. Malacol. Union Inc. A Rep.*, 1963 : 22.

—, 1979. *Economic Malacology with particular reference to Achatina fulica. In* : *Pulmonates*, 2B 1 : 150 pp. Academic Press, Inc., London, U.K., New York, USA.

—, 1982. *Malacologia*, **22** (1-2) : 489-493.

Mead, A.R. and Kemmerer, A.R., 1953. *Science*, **117** : 138-139.

Mead, A.R. and Kondo, Y., 1949. Giant African Snail (*Achatina fulica*) problem in Micronesia : Preliminary Report. Invert. Cons. Comm. for Micronesia, Pac. Sci. Board. Nat. Res. Council, 6 pp. (Typescript).

Meer Mohr. J.C. Van der., 1935. *Natur. und Volk*, **65** (2) : 62-67.

—, 1940. *Tropische Natuur*, **29** (6) : 145-148.

—, 1948. *Chronica naturae*, **104** (6) : 178-179.

—, 1949. *Treubia*, **20** (1) : 1-10.

—, 1957. *Indonesian Jour. Nat. Sci.*, **112** : 191-195.

Moreton, B.D., 1953. *Plant Path*, **2** : 49-51.

Muma, M.H., 1954. *Citrus Mag*, **16** (10) : 8-9.

Nair, M.R.G.K., Das, N.M. and Jacob, A., 1968. *Indian J. Ent.*, **30** : 50-60.

Nakata, S., 1960. *Proc. Hawaii. Entomol. Soc.*, **17** : 163.

Olson, F.J., 1973. The screening of candidate Molluscicides against the giant African snail, *Achatina fulica* Bowdich (Stylommatophora : Achatinidae). Thesis, University of Hawaii.

—, Economic Importance of the giant African snail, *Achatina fulica* Bowdich. (Unpublished).

Oughton, J., 1948. *Univ. Toronto Studies*, (Biol. Ser.) (57) : 126 pp.

Pangga, G.A., 1949. *Philipp. J. Agric.*, **14** (4) : 337-347.

Pappas, J.L. and Carman, G.E., 1955. *J. Econ. Ent.*, **48** (6) : 698-700.

Parker, T.J., Haswell, W.A. and Lavenstein, O., 1956. *A Text-book of Zoology*, Vol. I, MacMillan & Co. Ltd., London.

Pelseneer, P., 1935. *Essai d'ethologie zoologique d'apres l'etude des mollusques* Bruxelles, Palais des Academies, 662 pp.

Pemberton, C.E., 1938. *Hawaii. Planters' Rec.*, **42** (2) : 135-140.

Pemberton, C.E., Williams, F.X. and Sakimura, K., 1939. *Proc. Hawaii. Entomol Soc.*, **10** (2) : 187,190.

Pereira, H.F. and Goncalves, L.I., 1949. *O Biologico* (Sao Paulo), **15** (4) : 65-73.

Persing, C.O., 1945. *Calif. Citrograph*, **30** (4) : 94-95.

Peterson, G.D. Jr., 1957a. *J. Econ. Ent.*, **50** : 114.

—, 1957b. *Hilgardia*, **26** (16) : 643-658.

Philbrick, A.N., 1949. *The Times* (London) No. 51477 : 5.

Pillai, P.N., Radhakrishan and Jayarathnam, K., 1983. *In* : *Agricultural Entomology II* : 310-318. All India Scientific Writers' Society, New Delhi.

Pilsbry, H.A., 1919. *Bull. Amer. Mus. Nat. Hist.*, **50** (1) : 1-370.

Pilsbry, H.A. and Bequaert, J.C., 1927. *Bull. Amer. Mus. Nat. Hist.*, **53** (2) : 69-602.

Pradhan, S. and Srivastava, P.D., 1971. *Entomologists' Newslet*, **1** (9) : 62-63.

Rao, V.T., Ramdoss, P. and Rao, M.K., 1953. *Madras Agric. Jour.*, **40** (10) : 445-454.

Raut, S.K., 1978. *Bull. Zool. Surv. India*, **1** (3) : 243-246.

—, 1980. *Indian J. Anim. Health*, **19** (2) : 159-160.

—, 1981. *Progress in soil biology and ecology*, UAS Tech. Series, No. 37, 79-84.

—, 1982. *J. Zool. Soc. India*, **34** (1&2) : 7-12.

——, 1983a. *Proc. Workshop High Alt. Ent. & Wildl. Ecol., Zool. Surv. India*, Solan (H.P.) : 359-366.

——, 1983b. *Proc. Symp. Host. Environ., Zool. Surv. India*, Calcutta : 29-37.

——, 1983c. *J. Zool. Soc. India*, **3** (1&2) : 114-115.

——, 1986. *Env. and Ecol.*, **4** (1) : 187.

Raut, S.K. and Ghose, K.C., 1977. *Indian J. Anim. Health*, **16** (1) : 93-94.

——, 1979a. *Bull. Zool. Surv. India*, **2** (1) : 101-102.

——, 1979b. *Bull. Ent.*, **20** : 153-155.

——, 1983a. *Rec. Zool. Surv. India*, **80** (3&4) : 421-439.

——, 1983b. *Malacol. Rev.*, **16** : 95-96.

——, 1984. *Pestiferous Land snails of India*. Z.S.I. Tech. Monog. No. 11, 151 pp. Zoological Survey of India, Calcutta.

Ray, H.C. and Mukherjee, A., 1963. *Rec. Zool. Surv. India*, **61** (3&4) : 403-436.

Rees, W.J., 1950. *Loris*, **5** (4) : 159-161.

——, 1951. *Proc. Zool. Soc.* London, **120** (3) : 577-598.

Richards, C.S., 1968. *J. Protozool.*, **15** : 651-656.

Russel-Hunter, W.D., 1979. *A Life of Invertebrates*, MacMillan Publishing Co., Inc. N.Y.

Saxena, B.N. and Dubey, D.N., 1970. *Pesticides*, **4** (5) : 20-23.

Schmidt. H.A., 1955. *Arch. Molluskenk*, **84** : 167-177.

Schreurs, J., 1963. Investigations on the Biology, Ecology and Control of Giant African Snail in West New Guinea (Typed script, 18 pp.).

Shewell-Cooper, N.D.H., 1938. *The Gardner's Chronicle*, 3rd Ser. **103** (2677) : 272.

Singh, C. and Birat, R.B.S., 1969. *J. Bombay nat. Hist. Soc.*, **66** : 201-203.

Singh, U., Wadhwani, A.M. and Johri, B.M., 1983. *Dictionary of Economic Plants in India*, Indian Council of Agric. Res. New Delhi.

Somander, S.V.O., 1951. *Loris (Jour. Ceylon Wildlife)*, **5** (6) : 264-267.

South, F.W., 1923. *Malayan Agric. Jour.*, **11** (10) : 253.

——, 1926. *Malayan Agric. Jour.*, **14** (7) : 231-241.

Speyer, W., 1954. *Pflanzensch – utzdiehstes*, **6** (9) : 134-137.

Srivastava, P.D., 1966. *Indian J. Ent.*, **28** (3) : 412-413.

——, 1968a. *Indian J. Ent.*, **30** (3) : 217-219.

——, 1968b. *Indian J. Ent.*, **30** (3) : 240-241.

——, 1976. *Proc. natn. Acad, Sci.*, India, **46** (B) (I & II) : 60-64.

Srivastava, P.D. and Abbas, S.R. 1973. *Entomologists' Newslet.* **3** (5): 36.

——, 1983. In : *Agricultural Entomology II* : 340-355, All India Scientific Writers' Society, New Delhi.

Srivastava, P.D., Abbas, S.R. and Srivastava, Y.N., 1971. *Entomologists' Newslet.*, **1** (1) : 5.

Srivastava, P.D. and Srivastava, Y.N., 1967. *Experientia*, **23** : 776-777.

——, 1968. *The Veliger*, **10** (4) : 320-321.

Srivastava, P.D., Srivastava, Y.N. and Singh, J., 1968. *Indian J. Ent.*, **30** (2): 175-176.

Srivastava, P.D., Gupta, G.P. and Gautam, S.S.S., 1973. *Entomologists Newslet.*, **3** (7) : 47.

Srivastava, P.D., Gupta, G.P. and Srivastava, Y.N., 1987. *In*: *Recent Advances in Entomology* : 519-528. Gupta Prakashan, Kanpur : 519-528.

Srivastava, Y.N., Srivastava, P.D. and Doharey, K.L., 1975. *Entomologists' Newslet.*, **5** (10-11) : 50.

Storer, T.I. and Usinger, R.L., 1965. *General Zoology*, McGraw Hill Book Co., Inc., New York, London.

Stringer, A., 1946. *A. Rep. Agric. & Hort. Res. Sta.* Long Ashton, Bristol : 87-88.

Sturtevant, A.B. Jr., 1969. A Taxonomic study of the genus, *Aeromonas*. Thesis, University of Albama.

Subba Rao, N.V., 1975. *In : Dr. B.S. Chauhan Comm. Vol.*, 165-170.

Subba Rao, N.V. and Mitra, S.C., 1979. *Rec. Zool. Surv. India*, **75** : 1-37.

Symonds, B.V., 1975. *Plant Path.*, **24** : 1-9.

Thistle, A.D., 1953. *Hawaii. Board of Commissioners, Agric. and Forestry, A. Rep.*, 1953 : 28.

——, 1957. *Hawaii. Board of Commissioners, Agric. and Forestry, Biennial Rep.*, 1955/56 : 31-32, 36.

——, 1959. *Hawaii. Board of Commissioners, Agric. & Forestry, Binnial Rep.*, 1957-58 : 51-53, 59.

Thomas, D.C., 1948. *Ann. appl. Biol.*, **35** : 207-227.

Townes, H.K., 1946. U.S. Commer. Co. Econ. Surv., 1946, 53 pp. (Mimeographed).

Vagvolgyi, J., 1974. *In* : *First International Congress of Systematic and Evolutionary Biology*, Abstracts, University Colorado, Boulder : 8.

Van Cleave, H., 1953. *Smithsonian Inst. A. Rep.*, 1952 : 273-282.

Van Driest, J. Th., 1968. *Correspondentieblad Nederl. Malacol. Vereniging*, 127 : 1361-1362.

Van Zwaluwenburg, R.H., 1955. *Proc. Hawaii. Entomol Soc.*, **16** (1) : 1.

Verma, B.K., 1965. *Plant Protect. Bull.* (India), **14** : 21-22.

Voelker, J., 1959. *Mitt. Hamb. Zool. Mus. Inst.*, **57** : 37-78.

Vosburgh, F.G., 1950. *Natn. Geogr. Mag.*, **117** : 139-176.

Wadhi, S.R., 1982. In : *Agricultural Entomology* **1** : 158-171. All India Scientific Writers' Soc., New Delhi.

Weber, P.W., 1954. *Proc. Hawaii. Entomol. Soc.*, **15** (1) : 127-130.

Weel, P.B. Van., 1948a. *Chronica Naturae*, **104** (8/9) : 241-243.

—, 1948b. *Chronica Naturae*, **104** (10) : 278-280.

—, 1948c. *Chronica Naturae*, **104** (12) : 335-336.

Williams, F.X., 1951. *Bull. Mus. Comp. Zool.*, **105** (3) : 295-317.

Woglum, R.S., 1943. *Calif. Cult.*, **90** : 659.

Wolda, A., Zweep, A. and Schuitema, K.A., 1971. *Oecologia*, **7** : 361-381.

Wong, M., 1951. *Hawaii. Farm and Home*, **14** (8) : 26-28.

Wouters, L., 1970. *Pflanzenschutz - Nachr*, **23** : 165-169.

Wright, C.A., 1971. *Flukes and Snails*, George Allen & Unwin Ltd., London (U.K.).

INDEX